Computer Vision Using Local Binary Patterns

Computational Imaging and Vision

This comprehensive book series embraces state-of-the-art expository works and advanced research monographs on any aspect of this interdisciplinary field.

Topics covered by the series fall in the following four main categories:

- Imaging Systems and Image Processing
- Computer Vision and Image Understanding
- Visualization
- Applications of Imaging Technologies

Only monographs or multi-authored books that have a distinct subject area, that is where each chapter has been invited in order to fulfill this purpose, will be considered for the series.

Volume 40

For further volumes:
www.springer.com/series/5754

Matti Pietikäinen · Abdenour Hadid ·
Guoying Zhao · Timo Ahonen

Computer Vision Using Local Binary Patterns

 Springer

Matti Pietikäinen
Machine Vision Group
Department of Computer Science and
Engineering
University of Oulu
PO Box 4500
90014 Oulu
Finland
mkp@ee.oulu.fi

Abdenour Hadid
Machine Vision Group
Department of Computer Science and
Engineering
University of Oulu
PO Box 4500
90014 Oulu
Finland
hadid@ee.oulu.fi

Guoying Zhao
Machine Vision Group
Department of Computer Science and
Engineering
University of Oulu
PO Box 4500
90014 Oulu
Finland
gyzhao@ee.oulu.fi

Timo Ahonen
Nokia Research Center
Palo Alto, CA
USA
timo.ahonen@nokia.com

ISSN 1381-6446
ISBN 978-0-85729-747-1   e-ISBN 978-0-85729-748-8
DOI 10.1007/978-0-85729-748-8
Springer London Dordrecht Heidelberg New York

British Library Cataloguing in Publication Data
A catalogue record for this book is available from the British Library

Library of Congress Control Number: 2011932161

Mathematics Subject Classification: 68T45, 68H35, 68U10, 68T10, 97R40

Cover design: deblik

Printed on acid-free paper

Springer is part of Springer Science+Business Media (www.springer.com)

Preface

Humans receive the great majority of information about their environment through sight, and at least 50% of the human brain is dedicated to vision. Vision is also a key component for building artificial systems that can perceive and understand their environment. Computer vision is likely to change society in many ways; for example, it will improve the safety and security of people, it will help blind people see, and it will make human-computer interaction more natural. With computer vision it is possible to provide machines with an ability to understand their surroundings, control the quality of products in industrial processes, help diagnose diseases in medicine, recognize humans and their actions, and search for information from databases using image or video content.

Texture is an important characteristic of many types of images. It can be seen in images ranging from multispectral remotely sensed data to microscopic images. A textured area in an image can be characterized by a nonuniform or varying spatial distribution of intensity or color. The variation reflects some changes in the scene being imaged. For example, an image of mountainous terrain appears textured. In outdoor images, trees, bushes, grass, sky, lakes, roads, buildings etc. appear as different types of texture. The specific structure of the texture depends on the surface topography and albedo, the illumination of the surface, and the position and frequency response of the viewer. An X-ray of diseased tissue may appear textured due to the different absorption coefficients of healthy and diseased cells within the tissue.

Texture can play a key role in a wide variety of applications of computer vision. The traditional areas of application considered for texture analysis include biomedical image analysis, industrial inspection, analysis of satellite or aerial imagery, document image analysis, and texture synthesis for computer graphics or animation.

Texture analysis has been a topic of intensive research since the 1960s, and a wide variety of techniques for discriminating textures have been proposed. Most of the proposed methods have not been, however, capable to perform well enough for real-world textures and are computationally too complex to meet the real-time requirements of many applications. In recent years, very discriminative and computationally efficient local texture descriptors have been developed, such as local binary patterns (LBP), which has led to a significant progress in applying texture methods to various computer vision problems. The focus of the research has broadened from 2D textures to 3D textures and spatiotemporal (dynamic) textures.

With this progress the emerging application areas of texture analysis will also cover such modern fields as face analysis and biometrics, object recognition, motion analysis, recognition of actions, content-based retrieval from image or video databases, and visual speech recognition. This book provides an excellent overview how texture methods can be used for solving these kinds of problems, as well as more traditional applications. Especially the use of LBP in biomedical applications and biometric recognition systems has grown rapidly in recent years.

The local binary pattern (LBP) is a simple yet very efficient operator which labels the pixels of an image by thresholding the neighborhood of each pixel and considers the result as a binary number. The LBP method can be seen as a unifying approach to the traditionally divergent statistical and structural models of texture analysis. Perhaps the most important property of the LBP operator in real-world applications is its invariance against monotonic gray level changes caused, for example, by illumination variations. Another equally important is its computational simplicity, which makes it possible to analyze images in challenging real-time settings. LBP is also very flexible: it can be easily adapted to different types of problems and used together with other image descriptors.

The book is divided into five parts. Part I provides an introduction to the book contents and an in-depth description of the local binary pattern operator. A comprehensive survey of different variants of LBP is also presented. Part II deals with the analysis of still images using LBP operators. Applications in texture classification, segmentation, description of interest regions, content-based image retrieval and 3D recognition of textured surfaces are considered. The topic of Part III is motion analysis, with applications in dynamic texture recognition and segmentation, background modeling and detection of moving objects, and recognition of actions. Part IV deals with face analysis. The LBP operators are used for analyzing still images and image sequences. The specific application problem of visual speech recognition is presented in more detail. Finally, Part V provides an introduction to some related work by describing representative examples of using LBP in different applications, such as biometrics, visual inspection and biomedical applications, for example.

We would like to thank all co-authors of our LBP papers for their invaluable contributions to the contents of this book. First of all, special thanks to Timo Ojala and David Harwood who started LBP investigations in our group in fall 1992 during David Harwood's visit from the University of Maryland to Oulu. Since then Timo Ojala made many central contributions to LBP until 2002 when our very frequently cited paper was published in IEEE Transactions on Pattern Analysis and Machine Intelligence. Topi Mäenpää played also a very significant role in many developments of LBP. Other key contributors, in alphabetic order, include Jie Chen, Xiaoyi Feng, Yimo Guo, Chu He, Marko Heikkilä, Vili Kellokumpu, Stan Z. Li, Jiri Matas, Tomi Nurmela, Cordelia Schmid, Matti Taini, Valtteri Takala, and Markus Turtinen. We also thank the anonymous reviewers, whose constructive comments helped us improve the book.

Matlab and C codes of the basic LBP operators and some video demonstrations can be found from an accompanying website at www.cse.oulu.fi/MVG/LBP_Book. For a bibliography of LBP-related research and links to many papers, see www.cse. oulu.fi/MVG/LBP_Bibliography.

Oulu, Finland   Matti Pietikäinen
Oulu, Finland   Abdenour Hadid
Oulu, Finland   Guoying Zhao
Palo Alto, CA   Timo Ahonen

Contents

Abbreviations

1DHFLBP-TOP	One Dimensional Histogram Fourier LBP-TOP
2DHFLBP-TOP	Two Dimensional Histogram Fourier LBP-TOP
ALBP	Adaptive Local Binary Pattern
AM	Appearance-Motion
ARMA	Autoregressive and Moving Average
ASM	Active Shape Model
ASR	Audio only Speech Recognition
AVSR	Audio-Video Speech Recognition
BIC	Bayesian Intra/Extrapersonal Classifier
BLBP	Bayesian LBP
BSM	Binary Similarity Measures
CBIR	Content-Based Image Retrieval
CE	Capsule Endoscope
CLBP	Completed LBP
CNN-UM	Cellular Nonlinear Network-Universal Machine
Cohn-Kanade	A facial expression database
CRF	Conditional Random Field
CRIM	A video face database
CS-LBP	Center-Symmetric Local Binary Patterns
CT	Computed Tomography image
CTOP	Contrast from Three Orthogonal Planes
CUReT	A texture database
DFT	Discrete Fourier Transform
dLBP	Direction coded LBP
DLBP	Dominant Local Binary Patterns
DoG	Difference of Gaussians
DT	Dynamic Texture
DT-LBP	Decision Tree Local Binary Patterns
EBGM	Elastic Bunch Graph Matching
EBP	Elliptical Binary Patterns
EER	Equal Error Rate

E-GV-LBP	Effective Gabor Volume LBP
ELTP	Enlongated Ternary Patterns
EM	Expectation-Maximization
EPFDA	Ensemble of Piecewise Fisher Discriminant Analysis
EQP	Enlongated Quinary Patterns
EVLBP	Extended Volume Local Binary Patterns
FAR	False Acceptance Ratio
FCBF	Fast Correlation-Based Filtering
FDA	Fisher Discriminant Analysis
FERET	A face database
FLS	Filtering, Labeling and Statistic
FPLBP	Four-Patch Local Binary Patterns
FRGC	A face database
FSC	Fisher Separation Criteria
F-LBP	Fourier Local Binary Patterns
GFB	Gaussian Feature Bank
GMM	Gaussian Mixture Models
HCI	Human-Computer Interaction
HKLBP	Heat Kernel Local Binary Pattern
HLBP	Haar Local Binary Pattern
HMM	Hidden Markov Models
Honda/UCSD	A video face database
HOG	Histogram of Oriented Gradients
ILBP	Improved Local Binary Patterns
JAFFE	A facial expression database
KDCV	Kernel Discriminative Common Vectors
KTH-TIPS	Texture databases
LAB	Locally Assembled Binary Haar features
LABP	Local Absolute Binary Patterns
LBP	Local Binary Patterns
LBPV	Local Binary Pattern Variance
LBP/C	Joint distribution of LBP codes and a local Contrast measure
LBP-TOP	LBP from Three Orthogonal Planes
LBP-HF	Local Binary Pattern Histogram Fourier
LDA	Linear Discriminant Analysis
LDP	Local Derivative Patterns
LEP	Local Edge Patterns
LFW	The Labeled Faces in the Wild database
LGBP	Local Gabor Binary Patterns
LLBP	Local Line Binary Patterns
LP	Linear Programming
LPCA	Laplacian Principal Component Analysis
LPM	Local Pattern Model
LPP	Locality Preserving Projections
LPQ	Local Phase Quantization

LQP	Local Quinary Patterns
LTP	Local Ternary Patterns
MBP	Median Binary Patterns
MB-LBP	Multiscale Block Local Binary Pattern
MEI	Motion Energy Images
MHI	Motion History Images
MIR	Merger Importance Ratio
MLBP	Monogenic-LBP
MoBo	The CMU Motion of Body (MoBo) database
MR	Magnetic Resonance
MR8	A texture operator
MSF	Markov Stationary Features
MTL	Multi-Task Learning
NIR	Near-Infrared
OCLBP	Opponent Color Local Binary Patterns
Outex	A texture database
PCA	Principal Component Analysis
PLBP	Probabilistic LBP
PLS	Partial Least Squares
PPBTF	Pixel-Pattern-Based Texture Feature
RCC	Renal Cell Carcioma
SIFT	Scale Invariant Feature Transform
SILTP	Scale Invariant Local Ternary Pattern
SIMD	Single-Instruction Multiple-Data
SOM	Self-Organizing Map
SVM	Support Vector Machine
SVR	Support Vector Regression
S-LBP	Semantic Local Binary Patterns
tLBP	Transition coded LBP
TPLBP	Three-Patch Local Binary Patterns
VidTIMIT	An audio-video database
VLBP	Volume Local Binary Patterns
WLD	Weber Law Descriptor
XM2VTS	An audio-video database

Part I
Local Binary Pattern Operators

Chapter 1
Background

Visual detection and classification is of the utmost importance in several applications. Is there a human face in this image and if so, who is it? What is the person in this video doing? Has this photograph been taken inside or outside? Is there some defect in the textile in this image, or is it of acceptable quality? Does this microscope sample represent cancerous or healthy tissue?

To facilitate automated detection and classification in these types of questions, both good quality descriptors and strong classifiers are likely to be needed. In the appearance based description of images, a long way has been traveled since the pioneering work of Bela Julesz in [13], and good results have been reported in difficult visual classification tasks, such as texture classification, face recognition, and object categorization.

What makes the problem of visual detection and classification challenging is the great variability in real life images. Sources of this variability include viewpoint or lighting changes, background clutter, possible occlusion, non-rigid deformations, change of appearance over time, etc. Furthermore, image acquisition itself may present perturbations, like blur, due to the camera being out-of-focus, or noise.

Over the last few years, progress in the field of machine learning has manifested in learning based methods to cope with the variability in images. In practice, the system tries to learn the intra- and inter-class variability from, typically a very large set of, training examples. Despite the advances in machine learning, the maxim "garbage in, garbage out" still applies: if the features the machine learning algorithm is provided with do not convey the essential information for the application in question, good final results cannot be expected. In other words, good descriptors for image appearance are called for.

1.1 The Role of Texture in Computer Vision

Texture analysis has been a topic of intensive research since the 1960s, and a wide variety of techniques for discriminating textures have been proposed. A popular way

M. Pietikäinen et al., *Computer Vision Using Local Binary Patterns*,
Computational Imaging and Vision 40,
DOI 10.1007/978-0-85729-748-8_1, © Springer-Verlag London Limited 2011

is to divide them into four categories: statistical, geometrical, model-based and signal processing [36]. Among the most widely used traditional approaches are statistical methods based on co-occurrence matrices of second order gray level statistics [9] or first order statistics of local property values (difference histograms) [42], signal processing methods based on local linear transforms, multichannel Gabor filtering or wavelets [17, 22, 33], and model-based methods based on Markov random fields or fractals [5].

Most of the proposed methods have not been, however, capable to perform well enough for real-world textures and are computationally too complex to meet the real-time requirements of many computer vision applications. In recent years, very discriminative and computationally efficient local texture descriptors have been proposed, such as local binary patterns (LBP) [26, 28], which has led to a significant progress in applying texture methods to various computer vision problems. The focus of the research has broadened from 2D textures to 3D textures [6, 18, 37] and spatiotemporal (dynamic) textures [34, 35]. For a comprehensive description of recent progress in texture analysis, see the Handbook of Texture Analysis [23].

With this progress the application areas of texture analysis will also be covering such modern fields of computer vision as face and facial expression recognition, object recognition, background subtraction, visual speech recognition, and recognition of actions and gait.

1.2 Motivation and Background for LBP

The local binary pattern is a simple yet very efficient texture operator which labels the pixels of an image by thresholding the neighborhood of each pixel and considers the result as a binary number. The LBP method can be seen as a unifying approach to the traditionally divergent statistical and structural models of texture analysis. Perhaps the most important property of the LBP operator in real-world applications is its invariance against monotonic gray level changes caused, e.g., by illumination variations. Another equally important is its computational simplicity, which makes it possible to analyze images in challenging real-time settings.

The original local binary pattern operator, introduced by Ojala et al. [25, 26], was based on the assumption that texture has locally two complementary aspects, a pattern and its strength. The operator works in a 3×3 neighborhood, using the center value as a threshold. An LBP code is produced my multiplying the thresholded values with weights given by the corresponding pixels, and summing up the result. As the neighborhood consists of 8 pixels, a total of $2^8 = 256$ different labels can be obtained depending on the relative gray values of the center and the pixels in the neighborhood. The contrast measure (C) is obtained by subtracting the average of the gray levels below the center pixel from that of the gray levels above (or equal to) the center pixel. If all eight thresholded neighbors of the center pixel have the same value (0 or 1), the value of contrast is set to zero. The distributions of LBP codes, or two-dimensional distributions of LBP and local contrast (LBP/C),

Fig. 1.1   The original LBP

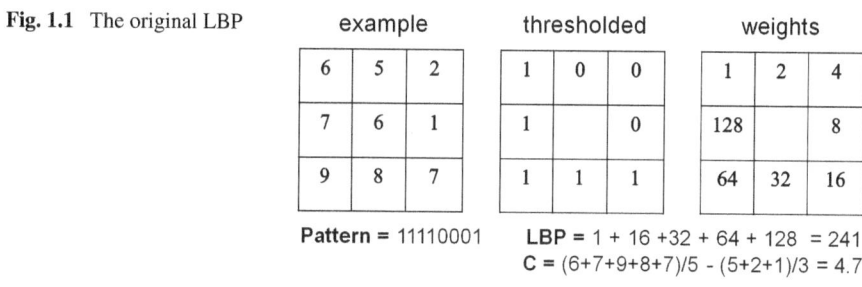

Pattern = 11110001

LBP = 1 + 16 +32 + 64 + 128 = 241

C = (6+7+9+8+7)/5 - (5+2+1)/3 = 4.7

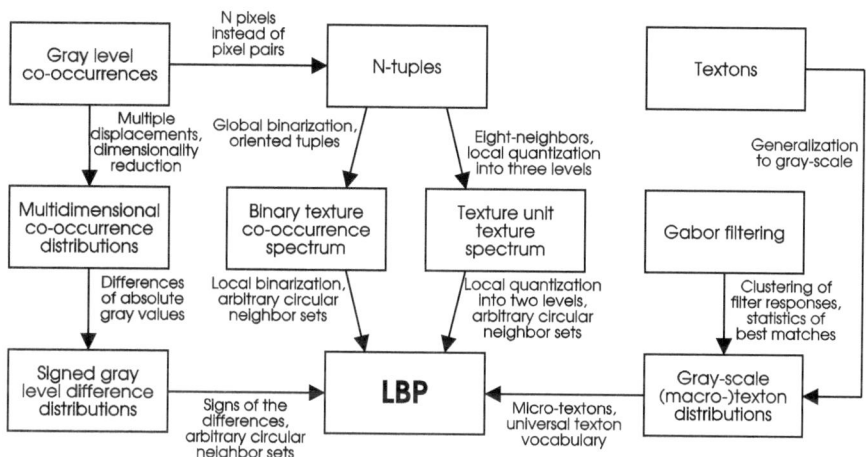

Fig. 1.2   Relation of LBP to earlier texture methods

are used as features in classification or segmentation. See Fig. 1.1 for an illustration of the basic LBP operator.

In its present form described in Chap. 2 the LBP is quite different from the basic version: the original version is extended to arbitrary circular neighborhoods and a number of extensions have been developed. The basic idea is however the same: the neighborhood of each pixel is binarized using thresholding.

The LBP is related to many well-known texture analysis operators as presented in Fig. 1.2 [19, 21]. The arrows represent the relations between different methods, and the texts beside the arrows summarize the main differences between them. And as shown in [2], LBP can also be seen as a combination of local derivative filter operators whose outputs are quantized by thresholding.

Due to its discriminative power and computational simplicity, the LBP texture operator has become a very popular approach in various applications. The great success of LBP in various texture analysis problems has shown that filter banks with large support areas are not necessary for high performance in texture classification, but operators defined for small neighborhoods such as LBP are usually adequate. A similar conclusion has been made for some other operators, see e.g. [38, 39]. The recent results demonstrate that an LBP-based approach has significant potential for many important tasks in computer vision which have not been earlier even regarded as texture problems. A proper exploitation of texture information could significantly increase the performance and reliability of many computer vision tasks and systems, helping make the technology inherently robust and simple to use in real-world applications.

1.3 A Brief History of LBP

The developments of LBP methodology can be divided into four main phases: (1) Introducing the basic LBP operator, (2) Developing extensions, generalizations and theoretical foundations of the operator, (3) Introducing methodology for face description based on LBPs, and (4) Spatiotemporal LBP operators for motion and activity analysis.

The basic LBP was developed during David Harwood's a few month's visit from the University of Maryland to Oulu in 1992. A starting point for the research was the idea that two-dimensional textures can be described by two complementary local measures: pattern and contrast. By separating pattern information from contrast, invariance to monotonic gray scale changes can be obtained. The use of whole feature distributions in texture classification, instead of e.g. means and variances, was also very rare in early 1990s. At that time the real value of LBP was not clear at all. The LBP was first published as a part of a comparative study of texture operators in the International Conference on Pattern Recognition conference (ICPR 1994) [25], and an extended version of it in Pattern Recognition journal [26]. The relation of LBP to the texture spectrum method proposed by Wang and He [41] was found during writing of the first paper on LBP. Years later it was also found that LBP developed for texture analysis is very similar to the census transform that was proposed at around the same time as LBP for computing visual correspondences in stereo matching [43]. The LBP and contrast operators introduced were later utilized for unsupervised texture segmentation [24], obtaining results which were clearly better than the state-of-the-art at that time. This showed the high potential of LBP and motivated for further research. Due to its computational simplicity the LBP was also used early in some applications like visual inspection, for example [31].

The development of a rotation-invariant LBP started in the late 1990s, and its first version was published in Pattern Recognition [30]. Another new development at that time was to investigate the relationship of the LBP to a method based on multidimensional gray scale difference histograms. This research was carried out together with Dr. Kimmo Valkealahti and Professor Erkki Oja from Helsinki University of Technology. As a result of this work, a method based on signed gray level differences was proposed [29], a simplification of which the LBP operator is. The signed difference operator used vector quantization to reduce the dimensionality of the feature space of multidimensional histograms and to form a one-dimensional texton histogram. Note that the texton-based texture operators later introduced e.g. in [39], utilizing image patch or filter response vectors followed by vector quantization, are closely related to this approach. These developments created theoretical basis for LBP and led to the development of the rotation-invariant multiscale LBP operator, the advanced version of which was published in IEEE Transactions on Pattern Analysis and Machine Intelligence in 2002 [27, 28]. After this the LBP became well known in the scientific community and its use in various applications increased significantly. The same article also introduced so-called "'uniform patterns'", which made a very simple rotation-invariant operator possible and have proven to be very important in reducing the feature vector length of the LBP needed in face recognition, for example. In early 2000s, an opponent color LBP was also proposed, and joint and separate use of color and texture in classification was studied [20]. The use of multiple LBP histograms in the classification of 3D textured surfaces was also considered [32]. Among the major developments of the spatial domain LBP operator since the mid 2000s were the center-symmetric LBP for interest region description [12] and LBP histogram Fourier features [4] for rotation-invariant texture description.

In 2004, a novel facial representation for face recognition based on LBP features was proposed. In this approach, the face image is divided into several regions from which the LBP features are extracted and concatenated into an enhanced feature vector to be used as a face descriptor [1]. A paper on this topic was later published in IEEE Transactions on Pattern Analysis and Machine Intelligence [3]. This approach has evolved to be a growing success. It has been adopted and further developed by a large number of research groups and companies around the world. The approach and its variants have been used to problems such as face recognition and authentication, face detection, facial expression recognition, gender classification and age estimation.

The use of LBP in motion analysis started with the development of a texture-based method for modeling the background and detecting moving objects in mid 2000s [10, 11]. Each pixel is modeled as a group of adaptive local binary pattern histograms that are calculated over a circular region around the pixel. The method was shown to be tolerant to illumination variations, the multimodality of the background, and the introduction or removal of background objects. The spatiotemporal VLBP and LBP-TOP proposed in 2007 created basis for many applications in motion and activity analysis [44], including facial expression recognition utilizing facial dynamics [44], face and gender recognition from video sequences [8], and recognition of actions and gait [14–16].

The development of different variants of spatial and spatiotemporal LBP has significantly increased in recent years, both in Oulu and elsewhere. Many of these will be briefly described or cited in the following chapters of this book.

1.4 Overview of the Book

The book is divided into five parts. Part I provides an introduction and in-depth description of the local binary pattern operator and its main variants. Part II deals with the analysis of still images using LBP operators in spatial domain. Applications in texture classification, segmentation, description of interest regions, content-based retrieval and 3D recognition are considered. The topic of Part III is motion analysis, with applications in dynamic textures, background modeling and recognition of actions. Part IV deals with face analysis. The LBP operators are used for analyzing both still images and image sequences. A specific application problem of visual speech recognition is presented in more detail. Finally, Part V describes briefly some interesting recent application studies using LBP.

A short introduction to the contents of different parts and chapters is given below.

Part I, composed of Chaps. 1–3, provides an introduction and in-depth description of the LBP operator and its main variants. Chapter 1 presents a background for texture-based approach to computer vision, motivations and brief history of the LBP operators, and an overview to the contents of the book. A detailed description of the LBP operators both in spatial and spatiotemporal domains is given in Chaps. 2 and 3.

Part II, divided into Chaps. 4–6, deals with applications of LBP in the analysis of still images. Most of the texture analysis research has been dealing with still images until recently. This is also the case with LBP methodology: during the first ten years of its existence almost all studies dealt with applications of LBP to single images. In this part, the use of LBP in important problems of texture classification, segmentation, description of interest regions, content-based image retrieval, and view-based recognition of 3D textured surfaces is considered.

Chapter 4 provides an introduction to the most common texture image test sets and overviews some texture classification experiments involving LBP descriptors. An unsupervised method for texture segmentation using LBP and contrast (LBP/C) distributions is also presented. This method has become very popular in the research community, and many variants of it have been proposed, for example for color-texture segmentation and segmentation of remotely sensed images. Chapter 5 introduces a method for interest region description using center-symmetric local binary patterns (CS-LBP). The CS-LBP descriptor combines the advantages of the well-known SIFT descriptor and the LBP operator. It performed better than SIFT in image matching experiments especially for image pairs having illumination variations. Chapter 6 considers two applications of LBP in spatial domain: Content-based image retrieval and recognition of 3D textured surfaces. Color and texture features are commonly used in retrieval, but usually they have been applied on full images. In the first part of this chapter two block based methods based on LBPs are presented which can significantly increase the retrieval performance. The second part describes a method for recognizing 3D textured surfaces using multiple LBP histograms as object models. Excellent results are obtained in view-based classification of the widely used CUReT texture database [7]. The method performed also well in the pixel-based classification of natural scene images.

Part III, consisting of Chaps. 7–9, considers applications of LBP in motion analysis. Motion is a fundamental property of an image sequence that carries information about temporal changes. While a still image contains only a snapshot of the scene at some time instant, an image sequence or video can capture temporal events and actions in the field of view. Motion also reveals the three-dimensional structure of the scene, which is not available from a single image frame. Motion plays a key role in many computer vision applications, including object detection and tracking, visual surveillance, human-computer interaction, video retrieval, 3D modeling, and video coding. The past research on motion analysis has been based on assumption that the scene is Lambertian, rigid and static. This kind of constraints greatly limit the applicability of motion analysis. Considering video sequences as dynamic textures allows to relax the constraints mentioned above [40]. The results on spatiotemporal (dynamic texture) extensions of LBP have shown very promising performance in various problems, including dynamic texture recognition and segmentation, facial expression recognition, lipreading, and activity and gait recognition.

In Chap. 7, recognition and segmentation of dynamic textures using spatiotemporal LBP operators are considered. Excellent classification performance is obtained for different test databases. The segmentation method extends the unsupervised segmentation method presented in Chap. 4 into spatiotemporal domain. Background subtraction, in which the moving objects are segmented from their background, is the first step in various applications of computer vision. Chapter 8 presents a robust texture-based method for modeling the background and detecting moving objects, obtaining state-of-the-art performance. The method has been successfully used in a multi-object tracking system, for example. Methods for analyzing humans and their actions from monocular or multi-view video data are required in applications such as visual surveillance, human-computer interaction, analysis of human activities in sports events or in psychological research, gait recognition (i.e. for identifying individuals in image sequences 'by the way they walk'), controlling video games on the basis of user's actions, and analyzing moving organs (e.g. a beating heart) in medical imaging. Chapter 9 introduces LBP-based approaches for action recognition. The methods perform very favorably compared to the state-of-the-art for test video sequences commonly used in the research community. A similar approach has also been successfully applied to gait recognition.

Part IV, composed of Chaps. 10–12, deals with applications of LBP methodology to face analysis problems. Detection and identification of human faces plays a key role in many emerging applications of computer vision, including biometric recognition systems, human-computer interfaces, smart environments, visual surveillance, and content-based image or video retrieval. Due to its importance, automatic face analysis which includes, for example, face detection and tracking, facial feature extraction, face recognition/verification, facial expression recognition and gender classification, has become one of the most active research topics in computer vision. Visual speech information plays an important role in speech recognition under noisy conditions or for listeners with hearing impairment. Therefore, automatic recognition of spoken phrases ("lipreading") is also an important research topic.

Chapter 10 considers face analysis using still images. It is explained how to easily derive efficient LBP based face descriptions which combine into a single feature vector the global shape and local texture of a facial image. The obtained representation is then applied to face and eye detection, face recognition, and facial expression recognition, yielding in excellent performance. In Chap. 11, spatiotemporal descriptors are applied to analyzing facial dynamics, with applications in facial expression, face and gender recognition from video sequences. Chapter 12 presents in more detail an approach for visual recognition of spoken phrases using LBP-TOP descriptors. The success of LBP in face description is due to the discriminative power and computational simplicity of the LBP operator, and the robustness of LBP to monotonic gray scale changes caused by, for example, illumination variations. The use of histograms as features also makes the LBP approach robust to face misalignment and pose variations. For these reasons, the LBP methodology has already attained an established position in face analysis research. This is attested by the increasing number of works which adopted a similar approach.

LBP has been used in a wide variety of different problems and applications around the world. Part V (Chap. 13) presents a brief introduction to some representative papers from different application areas.

References

1. Ahonen, T., Hadid, A., Pietikäinen, M.: Face recognition with local binary patterns. In: European Conference on Computer Vision. Lecture Notes in Computer Science, vol. 3021, pp. 469–481. Springer, Berlin (2004)
2. Ahonen, T., Pietikäinen, M.: Image description using joint distribution of filter bank responses. Pattern Recognit. Lett. **30**(4), 368–376 (2009)
3. Ahonen, T., Hadid, A., Pietikäinen, M.: Face description with local binary patterns: Application to face recognition. IEEE Trans. Pattern Anal. Mach. Intell. **28**(12), 2037–2041 (2006)
4. Ahonen, T., Matas, J., He, C., Pietikäinen, M.: Rotation invariant image description with local binary pattern histogram Fourier features. In: Scandinavian Conference on Image Analysis. Lecture Notes in Computer Science, vol. 5575, pp. 61–70. Springer, Berlin (2009)
5. Chellappa, R., Kashyap, R.L., Manjunath, B.S.: Model-based Texture Segmentation and Classification. In: Chen, C.H., Pau, L.F., Wang, P.S.P. (eds.) The Handbook of Pattern Recognition and Computer Vision, 2nd edn., pp. 249–282. World Scientific, Singapore (1998)
6. Cula, O.G., Dana, K.J.: Compact representation of bidirectional texture functions. In: Proc. IEEE Conference on Computer Vision and Pattern Recognition, vol. 1, pp. 1041–1047 (2001)
7. Dana, K.J., van Ginneken, B., Nayar, S.K., Koenderink, J.J.: Reflectance and texture of real-world surfaces. ACM Trans. Graph. **18**(1), 1–34 (1999)
8. Hadid, A., Pietikäinen, M.: Combining appearance and motion for face and gender recognition from videos. Pattern Recognit. **42**(11), 2818–2827 (2009)
9. Haralick, R.M., Dinstein, I., Shanmugaman, K.: Textural features for image classification. IEEE Trans. Syst. Man Cybern. **SMC-3**, 610–621 (1973)
10. Heikkilä, M., Pietikäinen, M.: A texture-based method for modeling the background and detecting moving objects. IEEE Trans. Pattern Anal. Mach. Intell. **28**(4), 657–662 (2006)
11. Heikkilä, M., Pietikäinen, M., Heikkilä, J.: A texture-based method for detecting moving objects. In: Proc. British Machine Vision Conference, pp. 187–196 (2004)
12. Heikkilä, M., Pietikäinen, M., Schmid, C.: Description of interest regions with local binary patterns. Pattern Recognit. **42**(3), 425–436 (2009)
13. Julesz, B.: Visual pattern discrimination. IRE Trans. Inf. Theory **8**(2), 84–92 (1962)
14. Kellokumpu, V., Zhao, G., Pietikäinen, M.: Human activity recognition using a dynamic texture based method. In: Proc. British Machine Vision Conference (2008)
15. Kellokumpu, V., Zhao, G., Pietikäinen, M.: Dynamic texture based gait recognition. In: Advances in Biometrics. Lecture Notes in Computer Science, vol. 5558, pp. 1000–1009. Springer, Berlin (2009)
16. Kellokumpu, V., Zhao, G., Pietikäinen, M.: Recognition of human actions using texture descriptors. Machine Vision and Applications (2011). doi: 10.1007/s00138-009-0233-8

17. Laws, K.I.: Texture energy measures. In: Proc. Image Understanding Workshop, pp. 47–51 (1979)
18. Leung, T., Malik, J.: Representing and recognizing the visual appearance of materials using three-dimensional textons. Int. J. Comput. Vis. **43**(1), 29–44 (2001)
19. Mäenpää, T.: The local binary pattern approach to texture analysis—extensions and applications. PhD thesis, Acta Universitatis Ouluensis C 187, University of Oulu (2003)
20. Mäenpää, T., Pietikäinen, M.: Classification with color and texture: Jointly or separately? Pattern Recognit. **37**, 1629–1640 (2004)
21. Mäenpää, T., Pietikäinen, M.: Texture Analysis with Local Binary Patterns. In: Chen, C.H., Wang, P.S.P. (eds.) Handbook of Pattern Recognition and Computer Vision, 3rd edn., pp. 197–216. World Scientific, Singapore (2005)
22. Manjunath, B.S., Ma, W.Y.: Texture features for browsing and retrieval of image data. IEEE Trans. Pattern Anal. Mach. Intell. **18**, 837–842 (1996)
23. Mirmehdi, M., Xie, X., Suri, J. (eds.): Handbook of Texture Analysis. Imperial College Press, London (2008)
24. Ojala, T., Pietikäinen, M.: Unsupervised texture segmentation using feature distributions. Pattern Recognit. **32**, 477–486 (1999)
25. Ojala, T., Pietikäinen, M., Harwood, D.: Performance evaluation of texture measures with classification based on Kullback discrimination of distributions. In: Proc. International Conference on Pattern Recognition, vol. 1, pp. 582–585 (1994)
26. Ojala, T., Pietikäinen, M., Harwood, D.: A comparative study of texture measures with classification based on feature distributions. Pattern Recognit. **29**(1), 51–59 (1996)
27. Ojala, T., Pietikäinen, M., Mäenpää, T.: Gray scale and rotation invariant texture classification with local binary patterns. In: European Conference on Computer Vision. Lecture Notes in Computer Science, vol. 1842, pp. 404–420. Springer, Berlin (2000)
28. Ojala, T., Pietikäinen, M., Mäenpää, T.: Multiresolution gray-scale and rotation invariant texture classification with local binary patterns. IEEE Trans. Pattern Anal. Mach. Intell. **24**(7), 971–987 (2002)
29. Ojala, T., Valkealahti, K., Oja, E., Pietikäinen, M.: Texture discrimination with multidimensional distributions of signed gray-level differences. Pattern Recognit. **34**(3), 727–739 (2001)
30. Pietikäinen, M., Ojala, T., Xu, Z.: Rotation-invariant texture classification using feature distributions. Pattern Recognit. **33**, 43–52 (2000)
31. Pietikäinen, M., Ojala, T., Nisula, J., Heikkinen, J.: Experiments with two industrial problems using texture classification based on feature distributions. In: Proc. SPIE Intelligent Robots and Computer Vision XIII: 3D Vision, Product Inspection, and Active Vision. Proc. SPIE, vol. 2354, pp. 197–204 (1994)
32. Pietikäinen, M., Nurmela, T., Mäenpää, T., Turtinen, M.: View-based recognition of real-world textures. Pattern Recognit. **37**(2), 313–323 (2004)
33. Randen, T., Husoy, J.H.: Filtering for texture classification: A comparative study. IEEE Trans. Pattern Anal. Mach. Intell. **21**(4), 291–310 (1999)
34. Saisan, P., Doretto, G., Wu, Y.N., Soatto, S.: Dynamic texture recognition. In: Proc. IEEE Conference on Computer Vision and Pattern Recognition, vol. 2, pp. 58–63 (2001)
35. Szummer, M., Picard, R.W.: Temporal texture modeling. In: Proc. IEEE International Conference on Image Processing, vol. 3, pp. 823–826 (1996)
36. Tuceryan, M., Jain, A.K.: Texture Analysis. In: Chen, C.H., Pau, L.F., Wang, P.S.P. (eds.) The Handbook of Pattern Recognition and Computer Vision, 2nd edn., pp. 207–248. World Scientific, Singapore (1998)
37. Varma, M., Zisserman, A.: Classifying images of materials: Achieving viewpoint and illumination independence. In: European Conference on Computer Vision. Lecture Notes in Computer Science, vol. 2352, pp. 255–271. Springer, Berlin (2002)
38. Varma, M., Zisserman, A.: Texture classification: Are filter banks necessary? In: Proc. IEEE Conference on Computer Vision and Pattern Recognition, vol. 2, pp. 691–698 (2003)
39. Varma, M., Zisserman, A.: A statistical approach to materials classification using image patch exemplars. IEEE Trans. Pattern Anal. Mach. Intell. **31**, 2032–2047 (2009)

40. Vidal, R., Ravichandran, A.: Optical flow estimation and segmentation of multiple moving dynamic textures. In: Proc. IEEE Conference on Computer Vision and Pattern Recognition, vol. 2, pp. 516–521 (2005)
41. Wang, L., He, D.C.: Texture classification using texture spectrum. Pattern Recognit. **23**, 905–910 (1990)
42. Weszka, J., Dyer, C., Rosenfeld, A.: A comparative study of texture measures for terrain classification. IEEE Trans. Syst. Man Cybern. **SMC-6**, 269–285 (1976)
43. Zabih, R., Woodfill, J.: Non-parametric local transforms for computing visual correspondence. In: European Conference on Computer Vision. Lecture Notes in Computer Science, vol. 801, pp. 151–158. Springer, Berlin (1994)
44. Zhao, G., Pietikäinen, M.: Dynamic texture recognition using local binary patterns with an application to facial expressions. IEEE Trans. Pattern Anal. Mach. Intell. **29**(6), 915–928 (2007)

Chapter 2
Local Binary Patterns for Still Images

The local binary pattern operator is an image operator which transforms an image into an array or image of integer labels describing small-scale appearance of the image. These labels or their statistics, most commonly the histogram, are then used for further image analysis. The most widely used versions of the operator are designed for monochrome still images but it has been extended also for color (multi channel) images as well as videos and volumetric data. This chapter covers the different versions of the actual LBP operator in spatial domain [42, 45, 53], while Chap. 3 deals with spatiotemporal LBP [88]. Parts II to IV of this book discuss how the labels are then used in different computer vision tasks.

2.1 Basic LBP

The basic local binary pattern operator, introduced by Ojala et al. [52], was based on the assumption that texture has locally two complementary aspects, a pattern and its strength. In that work, the LBP was proposed as a two-level version of the texture unit [74] to describe the local textural patterns.

The original version of the local binary pattern operator works in a 3×3 pixel block of an image. The pixels in this block are thresholded by its center pixel value, multiplied by powers of two and then summed to obtain a label for the center pixel. As the neighborhood consists of 8 pixels, a total of $2^8 = 256$ different labels can be obtained depending on the relative gray values of the center and the pixels in the neighborhood. See Fig. 1.1 for an illustration of the basic LBP operator. An example of an LBP image and histogram are shown in Fig. 2.1.

2.2 Derivation of the Generic LBP Operator

Several years after its original publication, the local binary pattern operator was presented in a more generic revised form by Ojala et al. [53]. In contrast to the basic

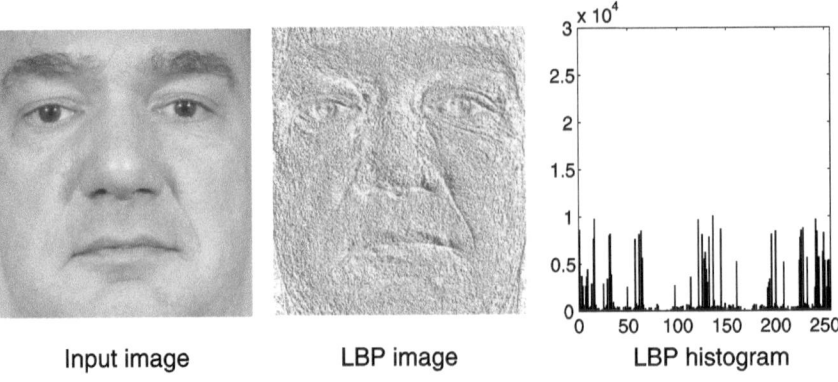

Input image   LBP image   LBP histogram

Fig. 2.1 Example of an input image, the corresponding LBP image and histogram

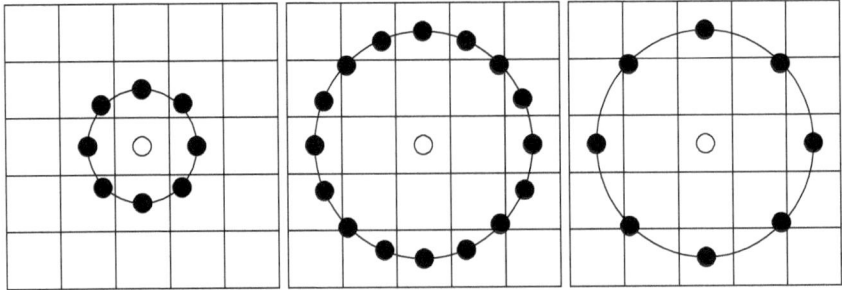

Fig. 2.2 The circular $(8, 1)$, $(16, 2)$ and $(8, 2)$ neighborhoods. The pixel values are bilinearly interpolated whenever the sampling point is not in the center of a pixel

LBP using 8 pixels in a 3×3 pixel block, this generic formulation of the operator puts no limitations to the size of the neighborhood or to the number of sampling points. The derivation of the generic LBP presented below follows that of [42, 45, 53].

Consider a monochrome image $I(x, y)$ and let g_c denote the gray level of an arbitrary pixel (x, y), i.e. $g_c = I(x, y)$.

Moreover, let g_p denote the gray value of a sampling point in an evenly spaced circular neighborhood of P sampling points and radius R around point (x, y):

$$g_p = I(x_p, y_p), \quad p = 0, \dots, P - 1 \quad \text{and} \tag{2.1}$$

$$x_p = x + R \cos(2\pi p / P), \tag{2.2}$$

$$y_p = y - R \sin(2\pi p / P). \tag{2.3}$$

See Fig. 2.2 for examples of local circular neighborhoods.

Assuming that the local texture of the image $I(x, y)$ is characterized by the joint distribution of gray values of $P + 1$ ($P > 0$) pixels:

$$T = t(g_c, g_0, g_1, \ldots, g_{P-1}). \tag{2.4}$$

Without loss of information, the center pixel value can be subtracted from the neighborhood:

$$T = t(g_c, g_0 - g_c, g_1 - g_c, \ldots, g_{P-1} - g_c). \tag{2.5}$$

In the next step the joint distribution is approximated by assuming the center pixel to be statistically independent of the differences, which allows for factorization of the distribution:

$$T \approx t(g_c)t(g_0 - g_c, g_1 - g_c, \ldots, g_{P-1} - g_c). \tag{2.6}$$

Now the first factor $t(g_c)$ is the intensity distribution over $I(x, y)$. From the point of view of analyzing local textural patterns, it contains no useful information. Instead the joint distribution of differences

$$t(g_0 - g_c, g_1 - g_c, \ldots, g_{P-1} - g_c) \tag{2.7}$$

can be used to model the local texture. However, reliable estimation of this multidimensional distribution from image data can be difficult. One solution to this problem, proposed by Ojala et al. in [54], is to apply vector quantization. They used learning vector quantization with a codebook of 384 codewords to reduce the dimensionality of the high dimensional feature space. The indices of the 384 codewords correspond to the 384 bins in the histogram. Thus, this powerful operator based on signed gray-level differences can be regarded as a texton operator, resembling some more recent methods based on image patch exemplars (e.g. [73]).

The learning vector quantization based approach still has certain unfortunate properties that make its use difficult. First, the differences $g_p - g_c$ are invariant to changes of the mean gray value of the image but not to other changes in gray levels. Second, in order to use it for texture classification the codebook must be trained similar to the other texton-based methods. In order to alleviate these challenges, only the signs of the differences are considered:

$$t(s(g_0 - g_c), s(g_1 - g_c), \ldots, s(g_{P-1} - g_c)), \tag{2.8}$$

where $s(z)$ is the thresholding (step) function

$$s(z) = \begin{cases} 1, & z \geq 0 \\ 0, & z < 0. \end{cases} \tag{2.9}$$

The generic local binary pattern operator is derived from this joint distribution. As in the case of basic LBP, it is obtained by summing the thresholded differences weighted by powers of two. The $\text{LBP}_{P,R}$ operator is defined as

$$\text{LBP}_{P,R}(x_c, y_c) = \sum_{p=0}^{P-1} s(g_p - g_c) 2^p. \tag{2.10}$$

In practice, Eq. 2.10 means that the signs of the differences in a neighborhood are interpreted as a P-bit binary number, resulting in 2^P distinct values for the LBP code. The local gray-scale distribution, i.e. texture, can thus be approximately described with a 2^P-bin discrete distribution of LBP codes:

$$T \approx t(\text{LBP}_{P,R}(x_c, y_c)). \tag{2.11}$$

In calculating the $\text{LBP}_{P,R}$ distribution (feature vector) for a given $N \times M$ image sample ($x_c \in \{0, \ldots, N-1\}$, $y_c \in \{0, \ldots, M-1\}$), the central part is only considered because a sufficiently large neighborhood cannot be used on the borders. The LBP code is calculated for each pixel in the cropped portion of the image, and the distribution of the codes is used as a feature vector, denoted by S:

$$S = t(\text{LBP}_{P,R}(x, y)),$$
$$x \in \{\lceil R \rceil, \ldots, N-1-\lceil R \rceil\}, y \in \{\lceil R \rceil, \ldots, M-1-\lceil R \rceil\}. \tag{2.12}$$

The original LBP (Fig. 1.1) is very similar to $\text{LBP}_{8,1}$, with two differences. First, the neighborhood in the general definition is indexed circularly, making it easier to derive rotation invariant texture descriptors. Second, the diagonal pixels in the 3×3 neighborhood are interpolated in $\text{LBP}_{8,1}$.

2.3 Mappings of the LBP Labels: Uniform Patterns

In many texture analysis applications it is desirable to have features that are invariant or robust to rotations of the input image. As the $\text{LBP}_{P,R}$ patterns are obtained by circularly sampling around the center pixel, rotation of the input image has two effects: each local neighborhood is rotated into other pixel location, and within each neighborhood, the sampling points on the circle surrounding the center point are rotated into a different orientation.

Another extension to the original operator uses so called *uniform patterns* [53]. For this, a uniformity measure of a pattern is used: U ("pattern") is the number of bitwise transitions from 0 to 1 or vice versa when the bit pattern is considered circular. A local binary pattern is called uniform if its uniformity measure is at most 2. For example, the patterns 00000000 (0 transitions), 01110000 (2 transitions) and 11001111 (2 transitions) are uniform whereas the patterns 11001001 (4 transitions) and 01010011 (6 transitions) are not. In uniform LBP mapping there is a separate output label for each uniform pattern and all the non-uniform patterns are assigned to a single label. Thus, the number of different output labels for mapping for patterns

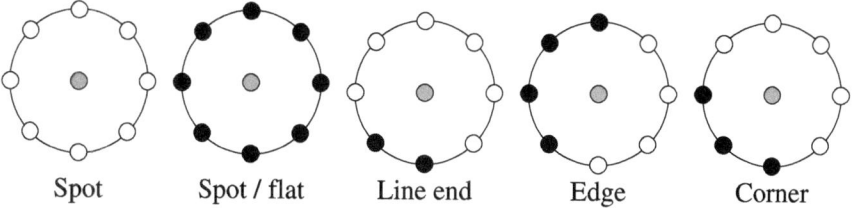

Fig. 2.3 Different texture primitives detected by the LBP

of P bits is $P(P - 1) + 3$. For instance, the uniform mapping produces 59 output labels for neighborhoods of 8 sampling points, and 243 labels for neighborhoods of 16 sampling points.

The reasons for omitting the non-uniform patterns are twofold. First, most of the local binary patterns in natural images are uniform. Ojala et al. noticed that in their experiments with texture images, uniform patterns account for a bit less than 90% of all patterns when using the $(8, 1)$ neighborhood and for around 70% in the $(16, 2)$ neighborhood. In experiments with facial images [4] it was found that 90.6% of the patterns in the $(8, 1)$ neighborhood and 85.2% of the patterns in the $(8, 2)$ neighborhood are uniform.

The second reason for considering uniform patterns is the statistical robustness. Using uniform patterns instead of all the possible patterns has produced better recognition results in many applications. On one hand, there are indications that uniform patterns themselves are more stable, i.e. less prone to noise and on the other hand, considering only uniform patterns makes the number of possible LBP labels significantly lower and reliable estimation of their distribution requires fewer samples.

The uniform patterns allows to see the LBP method as a unifying approach to the traditionally divergent statistical and structural models of texture analysis [45]. Each pixel is labeled with the code of the texture primitive that best matches the local neighborhood. Thus each LBP code can be regarded as a micro-texton. Local primitives detected by the LBP include spots, flat areas, edges, edge ends, curves and so on. Some examples are shown in Fig. 2.3 with the $LBP_{8,R}$ operator. In the figure, ones are represented as black circles, and zeros are white.

The combination of the structural and statistical approaches stems from the fact that the distribution of micro-textons can be seen as statistical placement rules. The LBP distribution therefore has both of the properties of a structural analysis method: texture primitives and placement rules. On the other hand, the distribution is just a statistic of a non-linearly filtered image, clearly making the method a statistical one. For these reasons, the LBP distribution can be successfully used in recognizing a wide variety of different textures, to which statistical and structural methods have normally been applied separately.

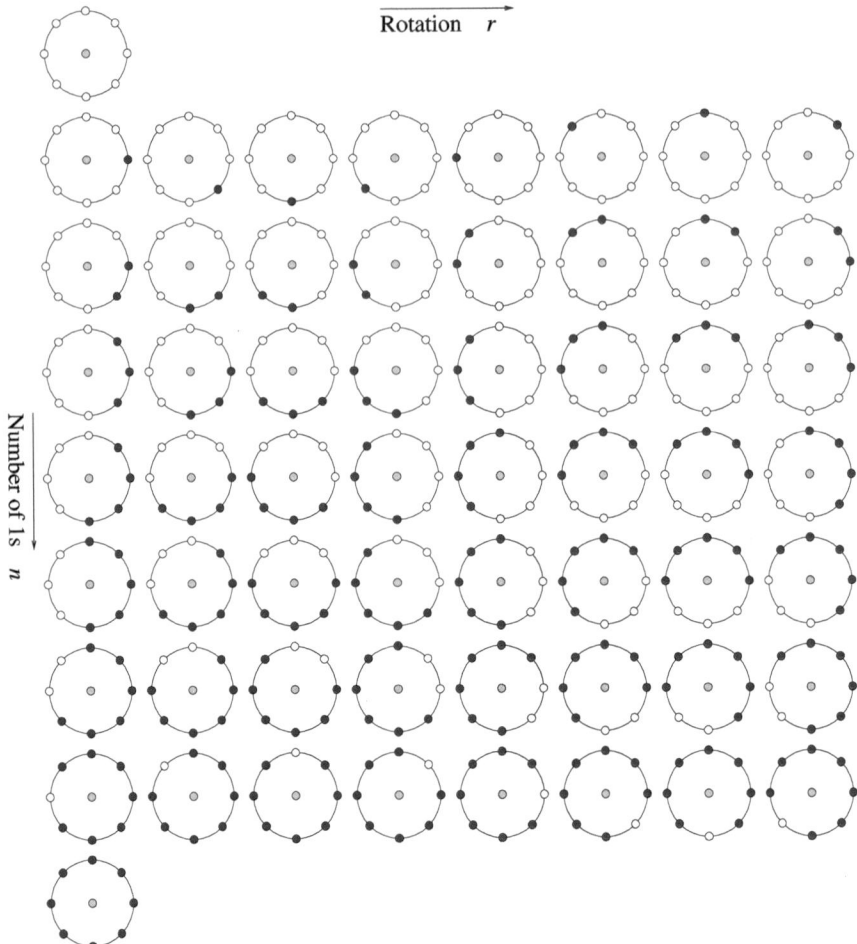

Fig. 2.4 The 58 different uniform patterns in $(8, R)$ neighborhood

2.4 Rotational Invariance

Let $U_P(n, r)$ denote a specific uniform LBP pattern. The pair (n, r) specifies a uniform pattern so that n is the number of 1-bits in the pattern (corresponds to row number in Fig. 2.4) and r is the rotation of the pattern (column number in Fig. 2.4) [6].

Now if the neighborhood has P sampling points, n gets values from 0 to $P + 1$, where $n = P + 1$ is the special label marking all the non-uniform patterns. Furthermore, when $1 \leq n \leq P - 1$, the rotation of the pattern is in the range $0 \leq r \leq P - 1$.

Let $I^{\alpha^\circ}(x, y)$ denote the rotation of image $I(x, y)$ by α degrees. Under this rotation, point (x, y) is rotated to location (x', y'). A circular sampling neighborhood on points $I(x, y)$ and $I^{\alpha^\circ}(x', y')$ also rotates by α°. See Fig. 2.5 [6].

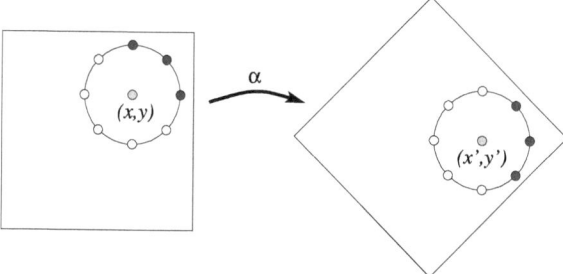

Fig. 2.5 Effect of image rotation on points in circular neighborhoods

If the rotations are limited to integer multiples of the angle between two sampling points, i.e. $\alpha = a\frac{360°}{P}, a = 0, 1, \ldots, P-1$, this rotates the sampling neighborhood by exactly a discrete steps. Therefore the uniform pattern $U_P(n, r)$ at point (x, y) is replaced by uniform pattern $U_P(n, r + a \bmod P)$ at point (x', y') of the rotated image.

From this observation, the original rotation invariant LBPs introduced in [53] and newer, histogram transformation based rotation invariant features described in [6] can be derived. These are discussed in the following.

2.4.1 Rotation Invariant LBP

As observed in the preceding discussion, rotations of a textured input image cause the LBP patterns to translate into a different location and to rotate about their origin. Computing the histogram of LBP codes normalizes for translation, and normalization for rotation is achieved by rotation invariant mapping. In this mapping, each LBP binary code is circularly rotated into its minimum value

$$\text{LBP}_{P,R}^{ri} = \min_i \text{ROR}(\text{LBP}_{P,R}, i), \tag{2.13}$$

where $\text{ROR}(x, i)$ denotes the circular bitwise right rotation of bit sequence x by i steps. For instance, 8-bit LBP codes 10000010b, 00101000b, and 00000101b all map to the minimum code 00000101b.

Omitting sampling artifacts, the histogram of $\text{LBP}_{P,R}^{ri}$ codes is invariant only to rotations of input image by angles $a\frac{360°}{P}, a = 0, 1, \ldots, P-1$. However classification experiments show that this descriptor is very robust to in-plane rotations of images by any angle.

2.4.2 Rotation Invariance Using Histogram Transformations

The rotation invariant LBP descriptor discussed above defined a mapping for individual LBP codes so that the histogram of the mapped codes is rotation invariant. In this section, a family of histogram transformations is presented that can be used to compute rotation invariant features from a uniform LBP histogram.

Consider the uniform LBP histograms $h_I(U_P(n, r))$. The histogram value h_I at bin $U_P(n, r)$ is the number of occurrences of uniform pattern $U_P(n, r)$ in image I.

If the image I is rotated by $\alpha = a\frac{360°}{P}$, this rotation of the input image causes a cyclic shift in the histogram along each of the rows,

$$h_{I_{\alpha°}}(U_P(n, r + a)) = h_I(U_P(n, r)). \tag{2.14}$$

For example, in the case of 8 neighbor LBP, when the input image is rotated by $45°$, the value from histogram bin $U_8(1, 0) = 000000001b$ moves to bin $U_8(1, 1) = 00000010b$, the value from bin $U_8(1, 1)$ to bin $U_8(1, 2)$, etc. Therefore, to achieve invariance to rotations of input image, features computed along the input histogram rows and are invariant to cyclic shifts can be used.

Discrete Fourier Transform is used to construct these features. Let $H(n, \cdot)$ be the DFT of nth row of the histogram $h_I(U_P(n, r))$, i.e.

$$H(n, u) = \sum_{r=0}^{P-1} h_I(U_P(n, r))e^{-i2\pi ur/P}. \tag{2.15}$$

In [6] it was shown that the Fourier magnitude spectrum

$$|H(n, u)| = \sqrt{H(n, u)\overline{H(n, u)}} \tag{2.16}$$

of the histogram rows results in features that are invariant to rotations of the input image.

Based on this property, an LBP-HF feature vector consisting of three LBP histogram values (all zeros, all ones, non-uniform) and Fourier magnitude spectrum values was defined. The feature vectors have the following form:

$$fv_{\text{LBP-HF}} = [|H(1, 0)|, \ldots, |H(1, P/2)|,$$

$$\ldots,$$

$$|H(P - 1, 0)|, \ldots, |H(P - 1, P/2)|,$$

$$h(U_P(0, 0)), h(U_P(P, 0)), h(U_P(P + 1, 0))]_{1 \times ((P-1)(P/2+1)+3)}.$$

It should also be noted that the Fourier magnitude spectrum contains rotation-invariant uniform pattern features LBP^{riu2} as a subset, since

$$|H(n, 0)| = \sum_{r=0}^{P-1} h_I(U_P(n, r)) = h_{\text{LBP}^{riu2}}(n). \tag{2.17}$$

An illustration of these features is in Fig. 2.6 [6].

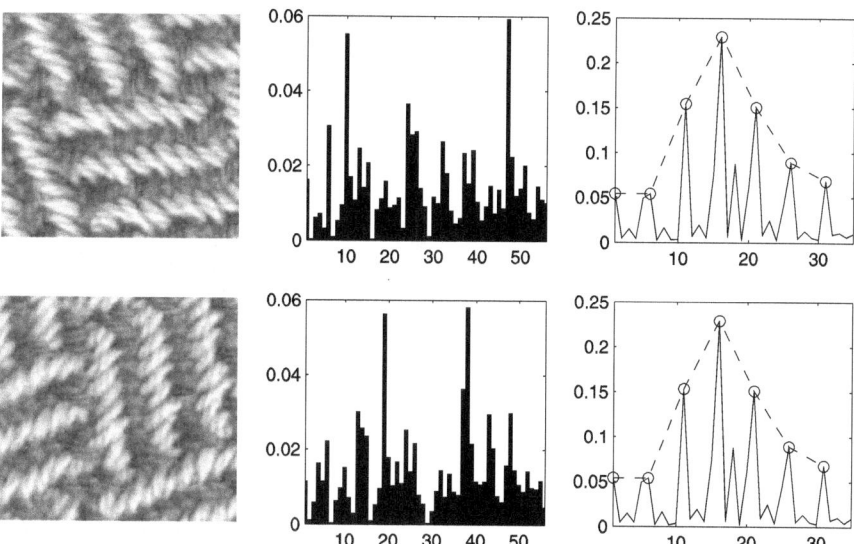

Fig. 2.6 *1st column*: Texture image at orientations $0°$ and $90°$. *2nd column*: bins 1–56 of the corresponding LBP^{u2} histograms. *3rd column*: Rotation invariant features $|H(n, u)|$, $1 \leq n \leq 7$, $0 \leq u \leq 5$, (*solid line*) and LBP^{riu2} (*circles, dashed line*). Note that the LBP^{u2} histograms for the two images are markedly different, but the $|H(n, u)|$ features are nearly equal

2.5 Complementary Contrast Measure

Contrast is a property of texture usually regarded as a very important cue for human vision, but the LBP operator by itself totally ignores the magnitude of gray level differences. In many applications, for example in industrial visual inspection, illumination can be accurately controlled. In such cases, a purely gray-scale invariant texture operator may waste useful information, and adding gray-scale dependent information may enhance the accuracy of the method. Furthermore, in applications such as image segmentation, gradual changes in illumination may not require the use of a gray-scale invariant method [42, 51].

In a more general view, texture is distinguished not only by texture patterns but also the strength of the patterns. Texture can thus be regarded as a two-dimensional phenomenon characterized by two orthogonal properties: spatial structure (patterns) and contrast (the strength of the patterns). Pattern information is independent of the gray scale, whereas contrast is not. On the other hand, contrast is not affected by rotation, but patterns are, by default. These two measures supplement each other in a very useful way. The LBP operator was originally designed just for this purpose: to complement a gray-scale dependent measure of the "amount" of texture. In [52], the joint distribution of LBP codes and a local contrast measure (LBP/C, see Fig. 1.1) is used as a texture descriptor.

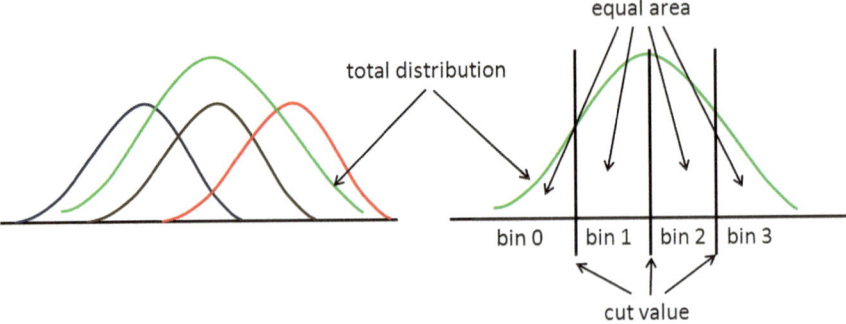

Fig. 2.7 Quantization of the feature space, when four bins are requested

Rotation invariant local contrast can be measured in a circularly symmetric neighbor set just like the LBP:

$$\text{VAR}_{P,R} = \frac{1}{P} \sum_{p=0}^{P-1} (g_p - \mu)^2, \quad \text{where } \mu = \frac{1}{P} \sum_{p=0}^{P-1} g_p. \tag{2.18}$$

$\text{VAR}_{P,R}$ is, by definition, invariant against shifts in the gray scale. Since contrast is measured locally, the measure can resist even intra-image illumination variation as long as the absolute gray value differences are not much affected. A rotation invariant description of texture in terms of texture patterns and their strength is obtained with the joint distribution of LBP and local variance, denoted as $\text{LBP}_{P_1,R_1}^{riu2}/\text{VAR}_{P_2,R_2}$. Typically, the neighborhood parameters are chosen so that $P_1 = P_2$ and $R_1 = R_2$, although nothing prevents one from choosing different values.

Variance measure has a continuous-valued output; hence, quantization of its feature space is needed. This can be done effectively by adding together feature distributions for every single model image in a total distribution, which is divided into B bins having an equal number of entries. Hence, the cut values of the bins of the histograms correspond to the $(100/B)$ percentile of the combined data. Deriving the cut values from the total distribution and allocating every bin the same amount of the combined data guarantees that the highest resolution of quantization is used where the number of entries is largest and vice versa. The number of bins used in the quantization of the feature space is of some importance as histograms with a too small number of bins fail to provide enough discriminative information about the distributions. On the other hand, since the distributions have a finite number of entries, a too large number of bins may lead to sparse and unstable histograms. As a rule of thumb, statistics literature often proposes that an average number of 10 entries per bin should be sufficient. In the experiments presented in this book, the value of B has been set so that this condition is satisfied. Figure 2.7 illustrates quantization of the feature space, when four bins are requested.

2.6 Non-parametric Classification Principle

In classification, the dissimilarity between a sample and a model LBP distribution is measured with a non-parametric statistical test. This approach has the advantage that no assumptions about the feature distributions need to be made. Originally, the statistical test chosen for this purpose was the cross-entropy principle [32, 52]. Later, Sokal and Rohlf [65] have called this measure the G statistic:

$$G(S, M) = 2 \sum_{b=1}^{B} S_b \log \frac{S_b}{M_b} = 2 \sum_{b=1}^{B} \left[S_b \log S_b - S_b \log M_b \right], \quad (2.19)$$

where S and M denote (discrete) sample and model distributions, respectively. S_b and M_b correspond to the probability of bin b in the sample and model distributions. B is the number of bins in the distributions [45].

For classification purposes, this measure can be simplified. First, the constant scaling factor 2 has no effect on the classification result. Furthermore, the term $\sum_{b=1}^{B} [S_b \log S_b]$ is constant for a given S, rendering it useless too. Thus the G statistic can be used in classification in a modified form:

$$L(S, M) = - \sum_{b=1}^{B} S_b \log M_b. \quad (2.20)$$

Model textures can be treated as random processes whose properties are captured by their LBP distributions. In a simple classification setting, each class is represented with a single model distribution M^i. Similarly, an unidentified sample texture can be described by the distribution S. L is a pseudo-metric that measures the likelihood that the sample S is from class i. The most likely class C of an unknown sample can thus be described by a simple nearest-neighbor rule:

$$C = \arg \min_i L(S, M^i). \quad (2.21)$$

Apart from a log-likelihood statistic, L can also be seen as a dissimilarity measure. Therefore, it can be used in conjunction with many classifiers, like the k-NN classifier, the self-organizing map (SOM) or the Support Vector Machine. The log-likelihood measure works well in many situations, but may be unstable with small sample sizes. The reason is that with small samples, the histogram is likely to contain many zeros, for which the logarithm is undefined. With small samples, the chi square distance usually works better [3]:

$$\chi^2(S, M) = \sum_{b=1}^{B} \frac{(S_b - M_b)^2}{S_b + M_b}. \quad (2.22)$$

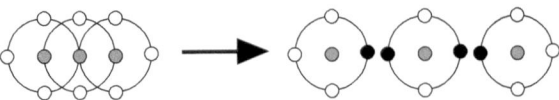

Fig. 2.8 Three adjacent LBP$_{4,R}$ neighborhoods and an impossible combination of codes. A *black disk* means the gray level of a sample is lower than that of the center

Almost equivalent accuracy can be achieved with the histogram intersection [67], with a significantly smaller computational overhead:

$$H(S, M) = \sum_{b=1}^{B} \min(S_b, M_b). \qquad (2.23)$$

2.7 Multiscale LBP

A significant limitation of the original LBP operator is its small spatial support area. Features calculated in a local 3×3 neighborhood cannot capture large-scale structures that may be the dominant features of some textures. However, adjacent LBP codes are not totally independent of each other. Figure 2.8 displays three adjacent four-bit LBP codes [42]. Assuming that the first bit in the leftmost code is zero, the third bit in the code to the right of it must be one. Similarly, the first bit in the code in the center and the third bit of the rightmost one must be either different or both equal to one. The right half of the figure shows an impossible combination of the codes. Each LBP code thus limits the set of possible codes adjacent to it, making the "effective area" of a single code actually slightly larger than 3×3 pixels. Nevertheless, the operator is not very robust against local changes in the texture, caused, for example, by varying viewpoints or illumination directions. An operator with a larger spatial support area is therefore often needed.

A straightforward way of enlarging the spatial support area is to combine the information provided by N LBP operators with varying P and R values. This way, each pixel in an image gets N different LBP codes. The most accurate information would be obtained by using the joint distribution of these codes. However, such a distribution would be overwhelmingly sparse with any reasonable image size. For example, the joint distribution of LBP$_{8,1}$, LBP$_{16,3}^{u2}$, and LBP$_{24,5}^{u2}$ would contain $256 \times 243 \times 555 \approx 3.5 \times 10^7$ bins. Therefore, only the marginal distributions of the different operators are considered even though the statistical independence of the outputs of the different LBP operators at a pixel cannot be warranted. For example, a feature histogram obtained by concatenating histograms produced by rotation-invariant uniform pattern operators at three scales (1, 3 and 5) is denoted as: LBP$_{8,1+16,3+24,5}^{riu2}$.

Fig. 2.9 LBP and CS-LBP
features for a neighborhood
of 8 pixels

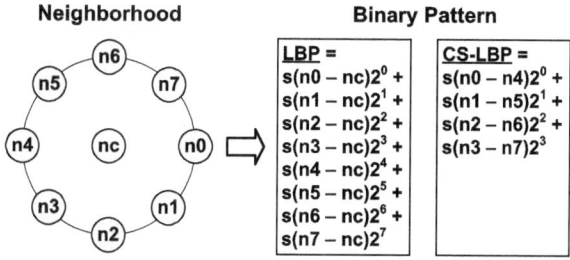

The aggregate dissimilarity between a sample and a model can be calculated as
a sum of the dissimilarities between the marginal distributions

$$L_N = \sum_{n=1}^{N} L(S^n, M^n),\qquad(2.24)$$

where S^n and M^n correspond to the sample and model distributions extracted by
the nth operator [53]. Of course, the chi square distance or histogram intersection
can also be used instead of the log-likelihood measure.

Even though the LBP codes at different radii are not statistically independent in
the typical case, using multi-resolution analysis often enhances the discriminative
power of the resulting features. With most applications, this straightforward way of
building a multi-scale LBP operator has resulted in very good accuracy.

2.8 Center-Symmetric LBP

Center-Symmetric Local Binary Patterns (CS-LBP) [23] were developed for interest region description. CS-LBP aims for smaller number of LBP labels to produce
shorter histograms that are better suited to be used in region descriptors. Also, CS-
LBP was designed to have higher stability in flat image regions.

In CS-LBP, pixel values are not compared to the center pixel but to the opposing
pixel symmetrically with respect to the center pixel. See Fig. 2.9 for an illustration
with eight neighbors [23].

Furthermore, to increase the operator's robustness in flat areas, the differences are
thresholded at a typically non-zero threshold T. CS-LBP operator is thus defined as

$$\text{CS-LBP}_{R,P,T}(x, y) = \sum_{p=0}^{(P/2)-1} s(g_p - g_{p+(P/2)} - T)2^p, \quad s(z) = \begin{cases} 1 & z \geq 0 \\ 0 & \text{otherwise}, \end{cases}$$

$$(2.25)$$

where n_i and $n_{i+(N/2)}$ correspond to the gray values of center-symmetric pairs of
pixels of N equally spaced pixels on a circle of radius R. It should be noticed that the
CS-LBP is closely related to gradient operator, because like some gradient operators
it considers gray level differences between pairs of opposite pixels in a neighborhood.

Based on the CS-LBP operator, Heikkilä et al. proposed a complete CS-LBP descriptor for interest regions. The steps of descriptor construction are summarized in the following. For more details, see [23] and Chap. 5.

1. Assuming that interest region with a known size and orientation has been detected, the region is normalized to a fixed size and orientation. In [23], 41×41 pixels was proposed as the size of the normalized region.
2. CS-LBP operator is applied to the normalized region.
3. The region is divided into cells. Authors suggest 3×3 or 4×4 Cartesian grids.
4. Histogram of the CS-LBP labels is constructed within each cell. To avoid boundary effects, bilinear interpolation is used to share the weight of each label between four nearest cells.
5. The histograms are concatenated to obtain the descriptor. The descriptor is then normalized to unit length, values above a pre-set threshold are clipped and finally the descriptor is re-normalized to unit length.

2.9 Other LBP Variants

The success of LBP methods in various computer vision problems and applications has inspired much new research on different variants. Due to its flexibility the LBP method can be easily modified to make it suitable for the needs of different types of problems. The basic LBP has also some problems that need to be addressed. Therefore, several extensions and modifications of LBP have been proposed with an aim to increase its robustness and discriminative power. In this section different variants are divided into such categories that describe their roles in feature extraction. Some of the variants could belong to more than one category, but in such cases only the most obvious category was chosen. A summary of the variants is presented in Table 2.1. The choice of a proper method for a given application depends on many factors, such as the discriminative power, computational efficiency, robustness to illumination and other variations, and the imaging system used. Therefore the LBP (and LBP with contrast) operators presented in the previous sections will usually provide a very good starting point when trying to find the optimal variant for a given application.

2.9.1 Preprocessing

In many applications, it is useful to preprocess the input image prior to LBP feature extraction. Especially multi-scale Gabor filtering and edge detection have been used for this purpose.

Gabor filtering has been widely used before LBP computation in face recognition. A motivation for this is that methods based on Gabor filtering and LBP provide complementary information: LBP captures small and fine details, while Gabor filters encode appearance information over a broader range of scales. For example,

Table 2.1 Summary of different LBP variants divided into such categories that describe their roles in feature extraction

Categories	LBP variants	Description	Ref.	Applications
Preprocessing	Local Gabor Binary Patterns (LGBP)	Gabor filtering before LBP computation	[85]	Face recognition
	Preprocessing chain	Gamma correction, DoG filtering, masking (optional), equalization of variation	[69]	Face recognition
	Local Edge Patterns (LEP)	Edge detection before LBP computation	[79]	Color texture retrieval
	Heat Kernel Local Binary Pattern (HKLBP)	Multiscale Heat kernel matrices are created before LBP computation	[37]	Face recognition
Neighborhood topology	Elliptical Binary Patterns (EBP)	Elliptical neighborhood is used	[38]	Face recognition
	Elongated Quinary Patterns (EQP)	Quinary encoding in elliptical neighborhood is used	[50]	Medical image analysis
	Local Line Binary Patterns (LLBP)	Lines in vertical and horizontal directions are used	[56]	Face recognition
	Three-Patch Local Binary Patterns (TPLBP)	Patch-based descriptor inspired by CS-LBP	[76]	Face analysis
	Four-Patch Local Binary Patterns (FPLBP)	Patch-based descriptor inspired by CS-LBP	[76]	Face analysis
Thresholding & encoding	Median Binary Patterns (MBP)	The median value within the neighborhood is used for thresholding	[19]	Texture classification
	Improved LBP (ILBP)	The mean of the local neighborhood is used for thresholding	[30]	Face detection
	Robust LBP	Differences thresholded at a non-zero threshold	[22]	Background subtraction
	Local Ternary Patterns (LTP)	Three values (1, 0 or -1) are used for encoding	[69]	Face recognition
	Elongated Quinary Patterns (EQP)	Five values ($-2, -1, 0, 1, 2$) are used for encoding	[50]	Medical image analysis
	Elongated Ternary Patterns (ELTP)	Three values are used for encoding	[49]	Medical image analysis
	Soft/Fuzzy Local Binary Patterns	Thresholding is replaced by a fuzzy membership function	[1, 29]	Texture analysis
	Probabilistic LBP	Thresholding is replaced by a probabilistic function	[68]	Face verification
	Scale Invariant Local Ternary Pattern (SILTP)	Extension of LTP to handle illumination variations	[40]	Background subtraction

Table 2.1 (Continued)

Categories	LBP variants	Description	Ref.	Applications
	Transition coded LBP (tLBP)	Encoding relation between neighboring pixels	[71]	Car detection
	Direction coded LBP (dLBP)	Related to CS-LBP, but uses also center pixel for encoding	[71]	Gender classification
	Centralized binary patterns (CBP)	Also related to CS-LBP, using center pixel for encoding	[14]	Facial expressions
	Semantic Local Binary Patterns (S-LBP)	Adding semantic consistency to LBP	[48]	Human detection
	Fourier Local Binary Patterns (F-LBP)	Adding semantic consistency to LBP	[48]	Human detection
	Local Derivative Patterns (LDP)	Encoding high order derivative patterns	[81]	Face recognition
	Bayesian Local Binary Patterns (BLBP)	Labeling is modeled as a probability and optimization process	[20]	Texture retrieval
Multiscale analysis	Gaussian filtering	Multiscale low-pass filtering before feature extraction	[43]	Texture classification
	Cellular automata	Compactly encoding several LBP operators at different scales	[43]	Texture classification
	Multiscale Block LBP (MB-LBP)	Compare average pixel values within small blocks	[41]	Face recognition
	Pyramid-based multi-structure LBP	Apply LBP on different layers of image pyramid	[21, 72]	Texture analysis
	Sparse multiscale LBP	Exploit the discriminative capacity of multiscale features	[59]	Texture/face recognition
	Multiresolution uniform patterns	Multiscale sampling points ordered according to sampling angle	[31]	Gait recognition
Handling rotation	Adaptive LBP (ALBP)	Directional statistical information is incorporated	[18]	Texture classification
	LBP variance (LBPV)	Build rotation variant LBP histogram and then apply a global matching	[17]	Texture classification
	Monogenic-LBP (M-LBP)	Integrates LBP with two other rotation-invariant measures	[84]	Texture classification
Handling color	Opponent color LBP	Each color channel and pairs of color channels are used	[44]	Texture analysis
	Separate color and texture	Texture and color features are computed separately	[44]	Texture analysis

Table 2.1 (Continued)

Categories	LBP variants	Description	Ref.	Applications
	Multiscale color LBPs	LBP values computed from different color channels	[92]	Object classes recognition
	Color vector LBP	Color LBP images computed considering pixels as color vectors	[58]	Color-texture classification
	3D histograms of LBP	LBP values computed from LBP images of three channels	[11]	Image indexing
Complementary descriptors	Completed LBP (CLBP)	Use local difference sign-magnitude transform	[16]	Texture classification
	Extended LBP	Encode both local gray level differences and ordinary LBP patterns	[25, 27]	Face recognition
	LBP and Gabor	Use of LBP and Gabor methods jointly	[70, 85]	Face recognition
	HOG-LBP	Combining LBP with the Histogram of Oriented Gradients operator	[75]	Human detection
	HOG-LBP-LTP	Combining LBP, HOG and LTP operators	[28]	Visual object detection
	CS-LBP descriptor	Combining the strengths of SIFT and LBP	[23]	Interest region description
	Haar-LBP	Combining ideas from Haar and LBP features	[60, 77]	Face detection
Feature selection and Learning	Dominant Local Binary Patterns (DLBP)	Make use of the most frequently occurred patterns of LBP	[39]	Texture classification
	Extended LBP	Analyzes the structure & occurrence probability of nonuniform patterns	[91]	Texture analysis
	LBP with Hamming distance	Nonuniform patterns incorporated into uniform patterns	[78]	Face recognition
	FSC-LBP	Fisher separation criterion is used to learn the most prominent pattern types	[15]	Texture classification
	Beam Search	Use beam search for selecting a subset of LBP patterns	[46]	Texture analysis
	Fast correlation-based filtering	Use of fast correlation-based filtering to select LBP patterns	[64]	Facial expression analysis

Table 2.1 (Continued)

Categories	LBP variants	Description	Ref.	Applications
	Symmetry patterns	Select patterns that contain high number of ones or zeros	[33]	Face recognition
	Decision tree LBP	Use decision tree algorithms to learn discriminative LBP-like patterns	[47]	Face recognition
	Boosting LBP bins	AdaBoost is used for learning discriminative LBP histogram bins	[61]	Facial expression recognition
	Boosting LBP regions	AdaBoost is used for selecting local regions and LBP settings	[82]	Face recognition
	Linear Discriminant Analysis	Use of LDA to project Multi-Scale LBP features to compact space	[7, 8]	Face recognition
	Ensemble of piecewise FDA	Construct ensemble of piecewise FDA for building compact LGBP feature space	[63]	Face recognition
	Kernel Discriminative Common vectors	Applied to Gabor wavelets and LBP features after PCA projection	[70]	Face recognition
	AdaBoost-LDA	Select most discriminative LBP features from a large pool of multiscale features	[24]	Face recognition
	Dual-Space LDA	Select discriminative LBP features	[86]	Face recognition
	Laplacian PCA	Select discriminative LBP features	[87]	Face recognition
	Partial Least Squares	PLS dimensionality reduction for selecting discriminative features	[28]	Visual object detector
	Locality Preserving Projections	Applied Locality Preserving Projections (LPP) on LBP features	[62]	Facial expression analysis
Other methods inspired by LBP	Weber Law Descriptor (WLD)	Codifies differential excitation and orientation components	[10]	Texture analysis
	Local Phase Quantization (LPQ)	Quantizing the Fourier transform phase in local neighborhoods	[55]	Texture classification
	GMM-based density estimator	Avoids the quantization errors of LBP	[34]	Texture analysis

Zhang et al. [85] proposed the extraction of LBP features from images obtained by filtering a facial image with 40 Gabor filters of different scales and orientations. The extracted features are called Local Gabor Binary Patterns (LGBP). Due to its high performance, the LGBP operator has been used as a reference method, together with the basic LBP method, in many recent face recognition studies. A downside of the method is the high dimensionality of the LGBP representation.

Tan and Triggs [69] developed a very effective preprocessing chain for compensating illumination variations in face images. It is composed of gamma correction, difference of Gaussian (DoG) filtering, masking (optional) and equalization of variation. This approach has been very successful in LBP-based face recognition under varying illumination conditions (see Chap. 10). When using it for the original LBP, the last step (i.e. equalization of variations) can be omitted due to LBP's invariance to monotonic gray scale changes.

In some studies edge detection has been used prior to LBP computation to enhance the gradient information. Yao and Chen [79] proposed local edge patterns (LEP) to be used with color features for color texture retrieval. In LEP, the Sobel edge detection and thresholding are used to find strong edges, and then LBP-like computation is used to derive the LEP patterns. In their method for shape localization Huang el al. [26] proposed an approach in which gradient magnitude images and original images are used to describe the local appearance pattern of each facial keypoint. A derivative-based LBP is used by applying LBP computation to the gradient magnitude image obtained by a Sobel operator. The Sobel-LBP later proposed by Zhao et al. [90] uses the same idea for facial image representation. First the Sobel edge detector is used and the LBPs are computed from the gradient magnitude images. They also applied Sobel-LBP on both the real and imaginary features of the Gabor filtered images.

Li et al. [37] proposed an approach based on capturing the intrinsic structural information of face appearances with multi-scale heat kernel matrices. Heat kernels perform well in characterizing the topological structural information of face appearance. Histograms of local binary patterns computed for non-overlapping blocks are then used for face description.

Also other types of preprocessing have been applied prior to LBP feature extraction. For example, computing LBPs from curvelet transformed images provided very promising performance in medical image analysis problems [35].

2.9.2 Neighborhood Topology

One important factor which makes the LBP approach so flexible to different types of problems is that the topology of the neighborhood from which the LBP features are computed can be different, depending on the needs of the given application.

The extraction of LBP features is usually done in a circular or square neighborhood. A circular neighborhood is important especially for rotation-invariant operators. However, in some applications, such as face recognition, rotation invariance is not required, but anisotropic information may be important. To exploit this, Liao and Chung used an elliptical neighborhood definition, calling their LBP variant an elliptical binary pattern (EBP). EBP, and EBP combined with a local gradient (contrast) measure, provided improved results in face recognition experiments compared to the ordinary LBP [38]. Nanni et al. investigated the use of different neighborhood topologies (circle, ellipse, parabola, hyperbola and Archimedean spiral) and encodings in their research on LBP variants for medical image analysis [50]. An operator using quinary encoding in an elliptic neighborhood (EQP) provided the best performance.

Petpon and Srisuk proposed to use lines in vertical and horizontal directions for LBP computations [56]. The Local Line Binary Pattern (LLBP) method then computes magnitudes by calculating the square root from the sum of the squared responses in these orthogonal directions.

Wolf et al. [76] considered different ways of using bit strings to encode the similarities between patches of pixels, which could capture complementary information to pixel-based descriptors. They proposed a Three-Patch LBP (TPLBP) and Four-Patch-LBP (FPLBP), which have borrowed some ideas from the Center-Symmetric LBP (CS-LBP) described earlier. For each pixel in TPLBP, for example, a $w \times w$ patch centered at the pixel and S additional patches distributed uniformly in a ring of radius r around it are considered. Then, the values for pairs of patches located on the circle at a specified distance apart are compared with those of the central patch. The value of a single bit is set according to which of the two patches is more similar to the central patch. The code produced will have S bits per pixel. In FPLBP, two rings centered on the pixel were used instead of one ring in TPLBP.

2.9.3 Thresholding and Encoding

Instead of using the value of the center pixel for thresholding in the local neighborhood, other techniques have also been considered. Hafiane et al. proposed Median Binary Pattern (MBP) operator by thresholding the local pixel values, including the center pixel, against the median (MBP) within the neighborhood [19]. The so-called Improved LBP, on the other hand, compares the values of the neighboring pixels against the mean gray level of the local neighborhood [13, 30]. A negative side is that the histograms for the methods using median or mean values for thresholding have 512 bins instead of 256 bins of the basic LBP. In fact, the use of the mean value of the local neighborhood was also considered, but not reported, when developing the original LBP in late 1992.

A drawback of the LBP method, as well as of all local descriptors that apply vector quantization, is that they are not robust in the sense that a small change in the input image would always cause a small change in the output. LBP may not work properly for noisy images or on flat image areas of constant gray level. This is due to the thresholding scheme of the operator.

In order to make the LBP more robust against these negligible changes in pixel values, the thresholding scheme of the operator was modified in [22] by replacing

Fig. 2.10 Local ternary
pattern operator

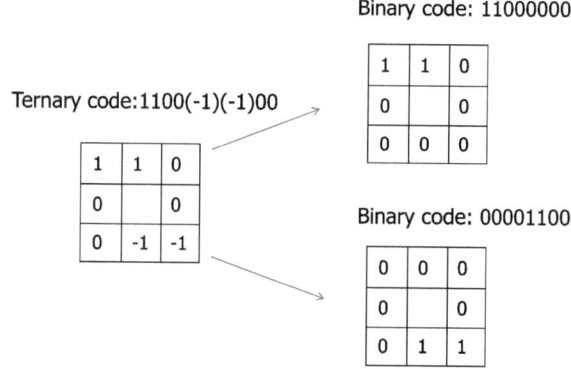

the term $s(g_p - g_c)$ in Eq. 2.10 with the term $s(g_p - g_c + a)$. The bigger the value of $|a|$ is, the bigger changes in pixel values are allowed without affecting the thresholding results. In order to retain the discriminative power of the LBP operator, a relatively small value should be used. In the experiments a was given a value of 3. An advantage of this robust LBP compared to the three-valued LBPs described below is that the feature vector length remains the same as in the ordinary LBP. A similar thresholding approach was also adopted to improve the robustness of CS-LBP as described in Sect. 2.8.

Tan and Triggs proposed a three-level operator called local ternary patterns (LTP) e.g. to deal with problems on near constant image areas [69]. In ternary encoding the difference between the center pixel and a neighboring pixel is encoded by three values (1, 0 or -1) according to a threshold T. The ternary pattern is divided into two binary patterns taking into account its positive and negative components. The histograms from these components computed over a region are then concatenated. Figure 2.10 depicts an example of splitting a ternary code into positive and negative codes. Note that LTP resembles the texture spectrum operator [74], which also used a three-valued output instead of two.

Nanni et al. [50] studied the effects of different encodings of the local grayscale differences, using binary (B), ternary (T) and a quinary (Q) encodings. In binary coding, the difference between a neighboring pixel and the center pixel is encoded by two values (0 and 1) like in LBP, in ternary encoding it is encoded by three values as in LTP, and in quinary encoding by five values $(-2, -1, 0, 1, 2)$ according to two thresholds (T1 and T2). A quinary code can be split into four binary LBP codes. In experiments with three different types of medical images the elongated quinary patterns (EQP) using elliptical neighborhoods provided the best overall performance. In their another study dealing with classification of pain states from facial expressions, the best results were obtained with elongated ternary patterns (ELTP) [49].

A soft three-valued LBP using fuzzy membership functions was proposed to improve the robustness in [1]. In soft LBP, one pixel typically contributes to more than one bin in the histogram. Fuzzy local binary patterns were also proposed by Iakovdis et al., with an application in ultrasound texture characterization [29]. A probabilistic

LBP (PLBP) was developed by Tan et al. [68], allowing to encode the magnitude of the difference between a neighboring pixel and the center pixel. A disadvantage of the fuzzy and probabilistic methods is their increased computational cost.

Liao et al. [40] noticed that adding a small offset value (T) for comparison in LTP is not invariant under scaling of intensity values. The intensity scale invariant property of a local comparison operator is very important for example in background modeling, because illumination variations, either global or local, often cause sudden changes of gray scale intensities of neighboring pixels simultaneously, which would approximately be a scale transform with a constant factor. Therefore, a Scale Invariant Local Ternary Pattern (SILTP) operator was developed for dealing with the gray scale intensity changes in complex background. Given any pixel location (x_c, y_c), SILTP encodes it as

$$\text{SILTP}_{N,R}^{\tau}(x_c, y_c) = \bigoplus_{k=0}^{N-1} s_{\tau}(I_c, I_k), \qquad (2.26)$$

where I_c is the gray intensity value of the center pixel, I_k are that of its N neighborhood pixels equally spaced on a circle of radius R, $\bigoplus$ denotes concatenation operator of binary strings, τ is a scale factor indicating the comparing range, and s_{τ} is a piecewise function defined as

$$s_{\tau}(I_c, I_k) = \begin{cases} 01, & \text{if } I_k > (1+\tau)I_c, \\ 10, & \text{if } I_k < (1-\tau)I_c, \\ 00, & \text{otherwise.} \end{cases} \qquad (2.27)$$

Since each comparison can result in one of three values, SILTP encodes it with two bits (with "11" undefined). The scale invariance of SILTP operator can be easily verified. The advantage of SILTP operator is in three fold. First, it is computationally efficient, which causes only one more comparison than LBP for each neighbor. Second, by introducing a tolerative range like LTP, the SILTP operator is robust to local image noise within a range. Especially in the shadowed area, the region is darker and contains more noise, in which SILTP is tolerable while local comparison results of LBP would be affected more. Finally, the scale invariance property makes SILTP robust to illumination changes. Assuming linear camera response, the SILTP feature is invariant if the illumination is suddenly changed from darker to brighter or vice versa. Besides, SILTP is robust when a soft shadow covers a background region, because the soft cast shadow reserves the background texture information but tends to be darker than the local background region with a scale factor.

A downside of the methods mentioned above using one or two thresholds is that the methods are not strictly invariant to local monotonic gray level changes as the original LBP. The feature vector lengths of these operators are also longer.

Trefny and Matas [71] proposed two new encoding schemes, which are complementary to the standard LBPs and also invariant to monotonic intensity transformations. The binary value transition coded LBP (tLBP) is composed of neighbor pixel comparisons in clockwise direction for all pixels except the central, encoding relation between neighboring pixels. Direction coded LBP (dLBP) is related to CS-LBP operator, but uses also center pixel information for encoding. Intensity variation along each of the four basic directions is coded into two bits. The first bit encodes whether the center pixel is an extrema and the second bit encodes whether difference of border pixels compared to the center pixel grows or falls. Experiments with face detection, car detection and gender recognition problems showed the efficiency of their approach. Another operator related to CS-LBP is centralized binary pattern (CBP) proposed by Fu and Wei [14] for facial expression recognition. CBP considers the contribution of the center pixel by comparing its value to the average of all nine pixels in the neighborhood, and encodes this bit with the largest weight.

Mu et al. developed LBP variants with an application in human detection in personal album [48]. They found that the original LBP does not suit so well for this problem due to its relatively high complexity and lack of semantic consistency. Therefore they proposed two variants of LBP: Semantic-LBP (S-LBP) and Fourier-LBP (F-LBP). First, a binarization of a pixel neighborhood is done on a color space like CIE-LAB. In S-LBP, several continuous "1" bits on the sampling circle form an arch, which can be represented with its principal direction and arch length. Non-uniform ones (with more than one arches) are abandoned. A two-dimensional histogram descriptor (arch angle vs. arch length) can be obtained for a given image region. In F-LBP, real valued color distance between the k-th samples and central pixel are computed and transformed into frequency domain. Low-frequency coefficients are then used to capture salient local structures around current pixel.

Inspired by LBP, higher order local derivative patterns (LDP) were proposed by Zhang et al., with applications in face recognition [81]. The basic LBP represents the first-order circular derivative pattern of images, a micropattern generated by the concatenation of the binary gradient directions as was shown in [2]. The higher order derivative patterns extracted by LDP will provide more detailed information, but may also be more sensitive to noise than in LBP.

Aiming at reducing the sensitivity of the image descriptor to illumination changes, a Bayesian LBP (BLBP) was developed by He et al. [20]. This operator is formulated in a Filtering, Labeling and Statistic (FLS) framework for texture descriptors. In the framework, the local labeling procedure, which is a part of many popular descriptors such as LBP and SIFT, can be modeled as a probability and optimization process. This enables the use of more reliable prior and likelihood information, and reduces the sensitivity to noise. The BLBP operator pursues a label image, when given the filtered vector image, by maximizing the joint probability of two images.

2.9.4 Multiscale Analysis

From a signal processing point of view, the sparse sampling used by multiscale LBP operators may not result in an adequate representation of the signal, resulting in aliasing effects [43]. Due to this some low-pass filtering would be needed to make the operator more robust. From the statistical point of view, however, even sparse sampling is acceptable provided that the number of samples is large enough. The sparse sampling is commonly used for example with the methods based on gray scale difference or co-occurrence statistics. Mäenpää and Pietikäinen proposed two alternative ways to multiscale analysis. In the first method Gaussian low-pass filters are used in collecting texture information from an larger area than the original single pixel. The filters and sampling position were designed to cope the neighborhood as well as possible while minimizing the redundant information. With this approach, the radii of the LBP operators used in the multiresolution version grow exponentially [43]. They also proposed another way of encoding arbitrarily large neighborhoods with cellular automata. It was used in compactly encoding even 12-scale LBP operators. A feature vector containing marginal distributions of LBP codes and cellular automation rules was used as a texture descriptor. In experiments, however, no clear improvement was obtained compared to the basic multi-scale approach.

Another extension of multiscale LBP operator is the multiscale block local binary pattern (MB-LBP) [41] which has gained popularity especially in facial image analysis. The key idea of MB-LBP is to compare average pixel values within small blocks instead of comparing pixel values. The operator always considers 8 neighbors, producing labels from 0 to 255. For instance, if the block size is 3×3 pixels, the corresponding MB-LBP operator compares the average gray value of the center block to the average values of the 8 neighboring blocks of the same size, thus the effective area of the operator is 9×9 pixels. Instead of the fixed uniform pattern mapping, MB-LBP has been proposed to be used with a mapping that is dynamically learned from a training data. In this mapping, the N most often occurring MB-LBP patterns receive labels $0, \ldots, N - 1$, and all the remaining patterns share a single label. The number of labels, and consequently the length of the MB-LBP histogram is a parameter the user can set.

A straightforward way for multiscale analysis is to utilize a pyramid of the input image computed at different resolutions, and then concatenate LBP distributions computed from different levels of the pyramid. In their research on contextual analysis of textured scene images Turtinen and Pietikäinen [72] combined this kind of idea with the original multiscale LBP approach: image patches at three different scales were resized to the same size and then LBP features were computed using LBPs with three different radii. Figure 2.11 illustrates the approach.

He et al. [21] developed a pyramid-based multistructure LBP for texture classification. It is obtained by executing the LBP on different layers of image pyramid, allowing to extract both micro and macro structures from textures. Five templates are used for creating the pyramid. The first one is a 2D Gaussian function used to smooth the image. Other four anisotropic filters are used to create anisotropic subimages of the pyramid in four directions. Good results are reported for the Outex textures, but the processing time is much higher than in the original LBP.

Raja and Gong proposed sparse multiscale local binary patterns to better exploit the discriminative capacity of multiscale features available [59]. A pairwise-coupled reformulation of LBP-type classification was used which involves selecting single-point features for each pair of classes to form compact, contextually-relevant multiscale predicates known as Multiscale Selected Local Binary Features (MSLBF).

Fig. 2.11 Multiscale feature extraction

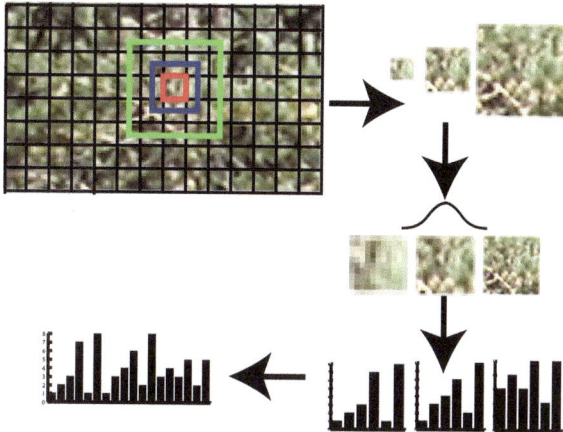

By the definition, uniform patterns are codes that consist of at most two bitwise transitions from 0 to 1 or vice versa when the binary string is considered circular. Therefore they can be considered as sectors on a sampling circle. When multiscale sampling points are ordered according to the sampling angle, they will also produce codes that satisfy the bit transition condition. Based on this observation, Kellokumpu et al. [31] proposed a new coding for multiresolution uniform patterns, obtaining improved results in gait recognition experiments.

2.9.5 Handling Rotation

LBPs have been used for rotation invariant texture recognition since late 1990s [57]. The most widely used version was proposed in [53] (Sect. 2.4.1), where the neighboring n binary bits around a pixel are clockwise rotated n times that a maximal number of the most significant bits is used to express this pixel.

Recently, in addition the method presented in Sect. 2.4.2, some other LBP variants for dealing with rotation have also been proposed.

Guo et al. developed an adaptive LBP (ALBP) [18] by incorporating the directional statistical information for rotation invariant texture classification. The directional statistical features, specifically the mean and standard deviation of the local absolute difference are extracted and used to improve the LBP classification efficiency. In addition, the least square estimation is used to adaptively minimize the local difference for more stable directional statistical features.

In [17], LBP variance (LBPV) was proposed as a rotation invariant descriptor. For LBPV there are three stages:

(1) putting the local contrast information into the one-dimensional LBP histogram; the variance $\text{VAR}_{P,R}$ was used as an adaptive weight to adjust the contribution

of the LBP code in histogram calculation. LBPV histogram is computed as:

$$\text{LBPV}_{P,R}(k) = \sum_{i=1}^{N} \sum_{j=1}^{M} w(\text{LBP}_{P,R}(i,j), k), \quad k \in [0, K], \quad (2.28)$$

where

$$w(\text{LBP}_{P,R}(i,j), k) = \begin{cases} \text{VAR}_{P,R}(i,j), & \text{LBP}_{P,R}(i,j) = k, \\ 0, & \text{otherwise;} \end{cases}$$

(2) learning the principal directions; the extracted LBPV features are used to estimate the principal orientations, and then the features are aligned to the principal
orientations, and
(3) determining the non-dominant patterns and thus by reducing them, feature dimension reduction was achieved.

Zhang et al. [84] proposed Monogenic-LBP (M-LBP), which integrates the traditional rotation-invariant LBP operator with two other rotation-invariant measures:
the local phase and local surface type computed by the first and second order Riesz
transforms, respectively. The local phase corresponds to a qualitative measure of
local structure (step, peak etc.), whereas the monogenic curvature tensor extracts
local surface type information. In experiments with CUReT textures the method
performed better than comparative methods (LBP, MR8, Joint), especially when the
training set was small and not comprehensive.

2.9.6 Handling Color

LBP operator was originally developed for monochrome images. There are many
possible ways for handling color with LBPs.

To describe color and texture jointly, opponent color LBP (OCLBP) [44] was
defined. In opponent color LBP, the operator is used on each color channel independently, and then for pairs of color channels so that the center pixel is taken from one
channel and the neighboring pixels from the other. Opposing pairs, such as R-G and
G-R are highly redundant, so either of them can be used in the analysis. In total, six
histograms (out of nine) are utilized (R, G, B, R-G, R-B, G-B), making the descriptor six times longer than the monochrome LBP histogram. Figure 2.12 illustrates
the three situations in which the center pixel is taken from the red channel [42].

The OCLBP descriptor fares well in comparison to other color texture descriptors. It has been later used successfully e.g. for face recognition by Chan et al. [7].
However, the authors of [44] do not recommend joint color and texture description as in their experiments "all joint color texture descriptors and all methods of
combining color and texture on a higher level are outperformed by either color or
gray-scale texture alone". This approach of handling color and texture separately
has been used in many recent studies.

Fig. 2.12 Opponent color
LBP for a *red center*. The
three planes illustrate color
channels

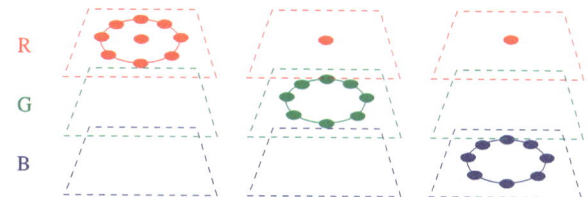

Instead of comparing the color components of pixels, Porebski et al. [58] considered color pixels represented by a vector when comparing neighboring pixels to the center pixel. Because there is no total order between vectors, they use a partial color order relation based on Euclidean distances for comparing their rank. As the result a single color LBP image is obtained instead of 6–9 provided by the OCLBP.

Another popular way is to apply the ordinary LBP to different color channels separately. Instead of the original R, G and B channels, other more discriminative and invariant color features derived from them can be used for LBP feature extraction as well. Along this line, Zhu et al. [92] proposed multiscale color LBPs for visual object classes recognition. Six operators were defined applying multiscale LBP on different types of channels and then concatenating the results together. From these the Hue-LBP (computed from the hue channel of the HSV color space), Opponent-LBP (computed over all three channels of the opponent color space) and nOpponent-LBP (computed over two channels of the normalized opponent space) provided the best performance on the well-known PASCAL Visual Object Classes Challenge 2007 benchmark (www.pascallin.ecs.soton.ac.uk/challenges/VOC/voc2007/).

Connah and Finlayson, on the other hand, used 3D histograms of LBP values computed from LBP images of three channels in their research on color constant image indexing [11]. They conclude that the good performance of the joint LBP histograms is a function of both their illumination invariance and their ability to encode additional information about the interaction between the color when using the three channels separately, whereas $10 \times 10 \times 10 = 1000$ bins are needed for a joint histogram.

2.9.7 Feature Selection and Learning

It has been shown by many studies that the dimensionality of the LBP distribution can be effectively decreased by reducing the number of neighboring pixels or by selecting a subset of bins available. In many cases a properly chosen subset of LBP patterns can perform better than the whole set of patterns.

Already the early studies on LBP indicated that in some problems considering only four neighbors of the center pixel (i.e. 16 bins) can provide almost as good results as eight neighbors (256 bins). Mäenpää et al. [46] showed that a major part of the discriminative power lies in a small properly selected subset of patterns. In addition to the uniform patterns (Sect. 2.3) they also considered a method based on beam search in which, starting from one, the size of the pattern set is iteratively increased up to a specified dimension D, and the best B pattern sets produced so far are always considered. In their experiments the method based on feature selection by beam search performed better than the whole set of patterns or the uniform patterns when classifying tilted textures and using nontilted samples for training. Thus the feature selection procedure was able to find those patterns that were able to survive the tilting best. In [64], Smith and Windeatt used the fast correlation-based filtering (FCBF) algorithm [80] to select the most discriminative LBP patterns. FCBF operates by repeatedly choosing the feature that is most correlated with a given class (e.g. person identity in case of face recognition), excluding those features already chosen or rejected, and rejecting any features that are more correlated with it than with the class. As a measure of correlation, the information-theoretic concept of symmetric uncertainty is used. When applied to the LBP features, FCBF reduced their number from 107,000 down to 120.

Lahdenoja et al. [33] defined a discrimination concept of the uniform LBP patterns called symmetry to reduce the feature vector length for LBP-based face description. Patterns are assigned different levels of symmetry based on the number of ones or zeros they contain. By definition these symmetry levels are rotation invariant. The patterns with a high level of symmetry were shown to be the most discriminative in experiments.

Liao et al. [39] introduced dominant local binary patterns (DLBP) which make use of the most frequently occurred patterns of LBP to improve the recognition accuracy compared to the original uniform patterns. The method has also rotation invariant characteristics. Zhou et al. considered that the LBP operator based on uniform patterns discards some important texture information and is sensitive to noise [91]. They proposed an extended LBP operator, which classifies and combines the nonuniform local patterns based on analyzing their structure and occurrence probability. Yang and Wang [78] also found that the nonuniform patterns contain useful information, incorporating these patterns into uniform patterns by minimizing the Hamming distance between them.

Guo et al. proposed a learning framework for image descriptor design [15]. The Fisher separation criterion (FSC) is used to learn the most reliable and robust dominant pattern types considering intraclass similarity and inter-class distance. Image structures are thus be described by a new FSC-based learning (FBL) encoding method. The learning framework includes three stages: (1) The learning stage. Determine most reliable dominant types for each class. Then, all the learnt dominant types of each class are merged and form the global dominant types for the whole database; (2) Extract global dominant types learnt in stage (1) of the training set; (3) Extract the global dominant types learnt in stage (1) of the testing set. Finally, features obtained in stages (2) and (3) are served as inputs to the classifier. FBL-LBP outperformed many other methods, including DLBP, in the experiments on three texture databases.

From the observation that LBP is equivalent to the application of a fixed binary decision tree, Maturana et al. [47] proposed a new method for learning discriminative LBP-like patterns from training data using decision tree induction algorithms. For each local image region, a binary decision tree is constructed from training data, thus obtaining an adaptive tree whose main branches are specially tuned to encode discriminative patterns in each region. Face recognition experiments on FERET and CAS-PEAL-R1 databases showed good performance compared to many traditional LBP-like approaches. Among the drawbacks of the proposed decision tree LBP (DT-LBP) is the high cost of constructing and storage of the decision trees especially when large pixel neighborhoods are used.

Boosting has become a very popular approach for feature selection. It has been widely adopted for LBP feature selection in various tasks e.g. 3D face recognition [36], face detection [83], gender classification [66] etc. AdaBoost is commonly used for selecting optimal LBP settings (such as the size and the location of local regions, the number of neighboring pixels etc.) or for selecting the most discriminative bins of an LBP histogram. For instance, Zhang et al. [82] used AdaBoost learning for selecting an optimal set for local regions and their weights for face recognition (see Chap. 10). Since then, many related approaches have been used at region level for LBP-based face analysis. Shan and Gritti [61], on the other hand, used AdaBoost for learning discriminative LBP histogram bins, with an application to facial expression recognition.

Another approach for deriving compact and discriminative LBP-based feature vectors consist of applying subspace methods for learning and projecting the LBP features from the original high-dimensional space into a lower dimensional space. For instance, Chan et al. used Linear Discriminant Analysis (LDA) to project high-dimensional Multi-Scale LBP features into a discriminant space [7, 8], yielding very promising results. To deal with the small sample size problem of LDA, Shan et al. [63] constructed ensemble of piecewise Fisher Discriminant Analysis (EPFDA) classifiers, each of which is designed based on one segment of the high-dimensional histogram of local Gabor binary pattern (LGBP) features. Their approach was shown to be more effective than applying LDA to high-dimensional holistic feature vectors.

Tan and Triggs [70] combined Gabor wavelets and LBP features and projected them to PCA space. Then, the Kernel Discriminative Common Vectors (KDCV) is applied to extract discriminant nonlinear compact features for face recognition. In [24], an AdaBoost-LDA learning algorithm was proposed to select the most discriminative LBP features from a large pool of multiscale features generated by shifting and scaling a subwindow over the image. Dual-Space LDA was also adopted to select discriminative LBP features in [86]. In another work, the authors applied Laplacian PCA (LPCA) for LBP feature selection and pointed out the superiority of LPCA over PCA and KPCA for feature selection [87]. In [28], the authors exploited the complementarity of three sets of features namely HOG features, local binary patterns (LBP) and local ternary patterns (LTP), and adopted Partial Least Squares (PLS) dimensionality reduction for selecting the most discriminative features, yielding fast and efficient visual object detector. In [62], Locality Preserving Projections (LPP) was applied on LBP features for embedding image sequences of facial expression from the high dimensional appearance feature space into a low dimensional manifold.

2.9.8 Complementary Descriptors

A current trend in the development of new effective local image and video descriptors is to combine the strengths of complementary descriptors. From the beginning the LBP operator was designed as a complementary measure of local image contrast. In many recent studies proposing new texture descriptors the role of the LBP contrast has not been considered when comparing LBP to the new descriptor. The use of LBP (or its simple robust version using a non-zero threshold [22], Sect. 2.9.3), can still be the method of choice for many applications, and should be considered when selecting a texture operator to be used. An interesting alternative for putting the local contrast into the one-dimensional LBP histogram was proposed by Guo et al. [17] (see Sect. 2.9.5).

In [16], a completed modeling of the LBP operator was proposed and an associated completed LBP (CLBP) scheme was developed for texture classification. The image local differences are decomposed into two complementary components: the signs and the magnitudes and two operators, CLBP-Sign (CLBP$_S$, also the original LBP) and CLBP-Magnitude (CLBP$_M$) were proposed to code them. As well, the center pixels represent the image gray level and they are converted into a binary code (CLBP$_C$) by global thresholding. The CLBP$_M$ and CLBP$_C$ were combined with CLBP$_S$ as complementary information to improve the texture classification. Earlier, a related Extended LBP (ELBP) was proposed which also encodes the local gray level differences in addition to the ordinary LBP computation [25, 27]. LBP codes are computed at multiple layers to encode the gray level differences between the center pixel and its neighbors.

Magnitude-LBP contains supplementary information to LBP. It was embedded to the histogram Fourier framework [89] and concatenated to LBPHF features as complementary descriptors to improve the description power for dealing with rotation variations.

In addition to applying LBP to Gabor-filtered face images, the use of LBP and Gabor methods jointly has provided excellent results in face recognition [70, 85]. The HOG-LBP, combining LBP with the Histogram of Oriented Gradients operator [12], has performed very well in human detection with partial occlusion handling [75]. Combining ideas from Haar and LBP features have given excellent results in accurate and illumination invariant face detection [60, 77]. A CS-LBP method for combining the strengths of SIFT and LBP in interest region description has also been developed [23] (Chap. 5).

2.9.9 Other Methods Inspired by LBP

LBP has also inspired the development of new effective local image descriptors.

The Weber Law Descriptor (WLD) is based on the fact that human perception of a pattern depends not only on the change of a stimulus (such as sound, lighting) but also on the original intensity of the stimulus [10]. Specifically, WLD consists of two components: differential excitation and orientation. The differential excitation component is a function of the ratio between two terms: one is the relative intensity differences of a current pixel against its neighbors; the other is the intensity of the current pixel. The orientation component is the gradient orientation of the current pixel. For a given image, the two components are used to construct a concatenated WLD histogram. Experimental results on texture analysis and face detection problems have provided excellent performance. Joint use of LBP and the excitation component of WLD descriptor in dynamic texture segmentation was considered in [9]. This indicates that this component could be useful in replacing the contrast measure of LBP also in other problems.

The local phase quantization (LPQ) descriptor is based on quantizing the Fourier transform phase in local neighborhoods [55]. The phase can be shown to be a blur invariant property under certain commonly fulfilled conditions. In texture analysis, histograms of LPQ labels computed within local regions are used as a texture descriptor similarly to the LBP methodology. The LPQ descriptor has received recently wide interest in blur-invariant face recognition [5]. LPQ is insensitive to image blurring, and it has proven to be a very efficient descriptor in face recognition from blurred as well as sharp images.

Lategahn et al. [34] developed a framework which filters a texture region by a set of filters and subsequently estimates the joint probability density functions by Gaussian mixture models (GMM). Using the oriented difference filters of the LBP method [2], they showed that this method avoids the quantization errors of LBP, obtaining better results than with the basic LBP. Additional performance improvement of the GMM-based density estimator was obtained when the elementary LBP difference filters were replaced by wavelet frame transform filter banks.

References

1. Ahonen, T., Pietikäinen, M.: Soft histograms for local binary patterns. In: Proc. Finnish Signal Processing Symposium, p. 4 (2007)
2. Ahonen, T., Pietikäinen, M.: Image description using joint distribution of filter bank responses. Pattern Recognit. Lett. **30**(4), 368–376 (2009)
3. Ahonen, T., Hadid, A., Pietikäinen, M.: Face recognition with local binary patterns. In: European Conference on Computer Vision. Lecture Notes in Computer Science, vol. 3021, pp. 469–481. Springer, Berlin (2004)
4. Ahonen, T., Hadid, A., Pietikäinen, M.: Face description with local binary patterns: Application to face recognition. IEEE Trans. Pattern Anal. Mach. Intell. **28**(12), 2037–2041 (2006)
5. Ahonen, T., Rahtu, E., Ojansivu, V., Heikkilä, J.: Recognition of blurred faces using local phase quantization. In: Proc. International Conference on Pattern Recognition, pp. 1–4 (2008)
6. Ahonen, T., Matas, J., He, C., Pietikäinen, M.: Rotation invariant image description with local binary pattern histogram Fourier features. In: Scandinavian Conference on Image Analysis. Lecture Notes in Computer Science, vol. 5575, pp. 61–70. Springer, Berlin (2009)
7. Chan, C.H., Kittler, J.V., Messer, K.: Multispectral local binary pattern histogram for component-based color face verification. In: Proc. IEEE Conference on Biometrics: Theory, Applications and Systems, pp. 1–7 (2007)
8. Chan, C.-H., Kittler, J., Messer, K.: Multi-scale local binary pattern histograms for face recognition. In: Proc. International Conference on Biometrics, pp. 809–818 (2007)

9. Chen, J., Zhao, G., Pietikäinen, M.: An improved local descriptor and threshold learning for unsupervised dynamic texture segmentation. In: Proc. ICCV Workshop on Machine Learning for Vision-based Motion Analysis, pp. 460–467 (2009)

10. Chen, J., Shan, S., He, C., Zhao, G., Pietikäinen, M., Chen, X., Gao, W.: WLD: A robust local image descriptor. IEEE Trans. Pattern Anal. Mach. Intell. **32**(9), 1705–1720 (2010)

11. Connah, D., Finlayson, G.D.: Using local binary pattern operators for colour constant image indexing. In: Proc. European Conference on Color in Graphics, Imaging, and Vision, p. 5 (2006)

12. Dalal, N., Triggs, B.: Histograms of oriented gradients for human detection. In: Proc. IEEE Conference on Computer Vision and Pattern Recognition, vol. 2, pp. 886–893 (2005)

13. Fröba, B., Ernst, A.: Face detection with the modified census transform. In: Proc. International Conference on Face and Gesture Recognition, pp. 91–96 (2004)

14. Fu, X., Wei, W.: Centralized binary patterns embedded with image Euclidean distance for facial expression recognition. In: Proc. International Conference on Natural Computation, vol. 4, pp. 115–119 (2008)

15. Guo, Y., Zhao, G., Pietikäinen, M., Xu, Z.: Descriptor learning based on Fisher separation criterion for texture classification. In: Proc. Asian Conference on Computer Vision, pp. 1491–1500 (2010)

16. Guo, Z.H., Zhang, L., Zhang, D.: A completed modeling of local binary pattern operator for texture classification. IEEE Trans. Image Process. **19**(6), 1657–1663 (2010)

17. Guo, Z.H., Zhang, L., Zhang, D.: Rotation invariant texture classification using LBP variance (LBPV) with global matching. Pattern Recognit. **43**(3), 706–719 (2010)

18. Guo, Z.H., Zhang, L., Zhang, D., Zhang, S.: Rotation invariant texture classification using adaptive LBP with directional statistical features. In: Proc. International Conference on Image Processing, pp. 285–288 (2010)

19. Hafiane, A., Seetharaman, G., Zavidovique, B.: Median binary pattern for texture classification. In: Proc. International Conference on Image Analysis and Recognition, pp. 387–398 (2007)

20. He, C., Ahonen, T., Pietikäinen, M.: A Bayesian local binary pattern texture descriptor. In: Proc. International Conference on Pattern Recognition, pp. 1–4 (2008)

21. He, Y., Sang, N., Gao, C.: Pyramid-based multi-structure local binary pattern for texture classification. In: Proc. Asian Conference on Computer Vision, vol. 3, pp. 1435–1446 (2010)

22. Heikkilä, M., Pietikäinen, M.: A texture-based method for modeling the background and detecting moving objects. IEEE Trans. Pattern Anal. Mach. Intell. **28**(4), 657–662 (2006)

23. Heikkilä, M., Pietikäinen, M., Schmid, C.: Description of interest regions with local binary patterns. Pattern Recognit. **42**(3), 425–436 (2009)

24. Ho An, K., Jin Chung, M.: Cognitive face analysis system for future interactive TV. IEEE Trans. Consum. Electron. **55**(4), 2271–2279 (2009)

25. Huang, D., Wang, Y., Wang, Y.: A robust method for near infrared face recognition based on extended local binary pattern. In: Advances in Visual Computing. Lecture Notes in Computer Science, vol. 4842, pp. 437–446. Springer, Berlin (2007)

26. Huang, X., Li, S.Z., Wang, Y.: Shape localization based on statistical method using extended local binary pattern. In: Proc. International Conference on Image and Graphics, pp. 184–187 (2004)

27. Huang, Y., Wang, Y., Tan, T.: Combining statistics of geometrical and correlative features for 3d face recognition. In: Proc. British Machine Vision Conference, pp. 879–888 (2006)

28. Hussain, S., Triggs, B.: Feature sets and dimensionality reduction for visual object detection. In: Proc. British Machine Vision Conference, pp. 112.1–112.10 (2010)

29. Iakovidis, D.K., Keramidas, E., Maroulis, D.: Fuzzy local binary patterns for ultrasound texture characterization. In: Proc. International Conference on Image Analysis and Recognition, pp. 750–759 (2008)

30. Jin, H., Liu, Q., Lu, H., Tong, X.: Face detection using improved LBP under Bayesian framework. In: Proc. International Conference on Image and Graphics, pp. 306–309 (2004)

31. Kellokumpu, V., Zhao, G., Pietikäinen, M.: Dynamic texture based gait recognition. In: Advances in Biometrics. Lecture Notes in Computer Science, vol. 5558, pp. 1000–1009. Springer, Berlin (2009)

32. Kullback, S.: Information Theory and Statistics. Dover, New York (1968)

33. Lahdenoja, O., Laiho, M., Paasio, A.: Reducing the feature vector length in local binary pattern based face recognition. In: Proc. International Conference on Image Processing, vol. 2, pp. 914–917 (2005)

34. Lategahn, H., Gross, S., Stehle, T., Aach, T.: Texture classification by modeling joint distributions of local patterns with Gaussian mixtures. IEEE Trans. Image Process. **19**, 1548–1557 (2010)

35. Li, B., Meng, M.Q.-H.: Texture analysis for ulcer detection in capsule endoscopy images. Image Vis. Comput. **27**, 1336–1342 (2009)

36. Li, S.Z., Zhao, C., Zhu, X., Lei, Z.: Learning to fuse 3D+2D based face recognition at both feature and decision levels. In: Proc. IEEE International Workshop on Analysis and Modeling of Faces and Gestures, pp. 44–54 (2005)

37. Li, X., Hu, W., Zhang, Z., Wang, H.: Heat kernel based local binary pattern for face representation. IEEE Signal Process. Lett. **17**, 308–311 (2010)

38. Liao, S., Chung, A.C.S.: Face recognition by using enlongated local binary patterns with average maximum distance gradient magnitude. In: Computer Vision—ACCV 2007. Lecture Notes in Computer Science, vol. 4844, pp. 672–679. Springer, Berlin (2007)

39. Liao, S., Law, M., Chung, C.S.: Dominant local binary patterns for texture classification. IEEE Trans. Image Process. **18**, 1107–1118 (2009)

40. Liao, S., Zhao, G., Kellokumpu, V., Pietikäinen, M., Li, S.Z.: Modeling pixel process with scale invariant local patterns for background subtraction in complex scenes. In: Proc. IEEE Conference on Computer Vision and Pattern Recognition, p. 8 (2010)

41. Liao, S., Zhu, X., Lei, Z., Zhang, L., Li, S.Z.: Learning multi-scale block local binary patterns for face recognition. In: Proc. International Conference on Biometrics, pp. 828–837 (2007)

42. Mäenpää, T.: The local binary pattern approach to texture analysis—extensions and applications. PhD thesis, Acta Universitatis Ouluensis C 187, University of Oulu (2003)

43. Mäenpää, T., Pietikäinen, M.: Multi-scale binary patterns for texture analysis. In: Scandinavian Conference on Image Analysis. Lecture Notes in Computer Science, vol. 2749, pp. 885–892. Springer, Berlin (2003)

44. Mäenpää, T., Pietikäinen, M.: Classification with color and texture: Jointly or separately? Pattern Recognit. **37**, 1629–1640 (2004)

45. Mäenpää, T., Pietikäinen, M.: Texture analysis with local binary patterns. In: Chen, C.H., Wang, P.S.P. (eds.) Handbook of Pattern Recognition and Computer Vision, 3rd edn., pp. 197–216. World Scientific, Singapore (2005)

46. Mäenpää, T., Ojala, T., Pietikäinen, M., Soriano, M.: Robust texture classification by subsets of local binary patterns. In: Proc. 15th International Conference on Pattern Recognition, vol. 3, pp. 947–950 (2000)

47. Maturana, D., Soto, A., Mery, D.: Face recognition with decision tree-based local binary patterns. In: Proc. Asian Conference on Computer Vision, 2010

48. Mu, Y.D., Yan, S.C., Liu, Y., Huang, T., Zhou, B.F.: Discriminative local binary patterns for human detection in personal album. In: Proc. IEEE Conference on Computer Vision and Pattern Recognition, pp. 1–8 (2008)

49. Nanni, L., Brahnam, S., Lumini, A.: A local approach based on a local binary patterns variant texture descriptor. Expert Syst. Appl. **37**, 7888–7894 (2010)

50. Nanni, L., Lumini, A., Brahnam, S.: Local binary patterns variants as texture descriptors for medical image analysis. Artif. Intell. Med. **49**, 117–125 (2010)

51. Ojala, T., Pietikäinen, M.: Unsupervised texture segmentation using feature distributions. Pattern Recognit. **32**, 477–486 (1999)

52. Ojala, T., Pietikäinen, M., Harwood, D.: A comparative study of texture measures with classification based on feature distributions. Pattern Recognit. **29**(1), 51–59 (1996)

53. Ojala, T., Pietikäinen, M., Mäenpää, T.: Multiresolution gray-scale and rotation invariant texture classification with local binary patterns. IEEE Trans. Pattern Anal. Mach. Intell. **24**(7), 971–987 (2002)
54. Ojala, T., Valkealahti, K., Oja, E., Pietikäinen, M.: Texture discrimination with multidimensional distributions of signed gray-level differences. Pattern Recognit. **34**(3), 727–739 (2001)
55. Ojansivu, V., Heikkilä, J.: Blur insensitive texture classification using local phase quantization. In: Proc. International Conference on Image and Signal Processing, pp. 236–243 (2008)
56. Petpon, A., Srisuk, S.: Face recognition with local line binary pattern. In: Proc. International Conference on Image and Graphics, pp. 533–539 (2009)
57. Pietikäinen, M., Ojala, T., Xu, Z.: Rotation-invariant texture classification using feature distributions. Pattern Recognit. **33**, 43–52 (2000)
58. Porebski, A., Vandenbroucke, N., Macaire, L.: Haralick feature extraction from LBP images for color texture classification. In: Proc. Workshop on Image Processing Theory, Tools and Applications, pp. 1–8 (2008)
59. Raja, Y., Gong, S.: Sparse multiscale local binary patterns. In: Proc. British Machine Vision Conference, 2006
60. Roy, A., Marcel, S.: Haar local binary pattern feature for fast illumination invariant face detection. In: Proc. British Machine Vision Conference, 2009
61. Shan, C., Gritti, T.: Learning discriminative LBP-histogram bins for facial expression recognition. In: Proc. British Machine Vision Conference, p. 10 (2008)
62. Shan, C., Gong, S., Mcowan, P.: Appearance manifold of facial expression. In: Proc. IEEE ICCV Workshop on Human-Computer Interaction (HCI), pp. 221–230 (2005)
63. Shan, S., Zhang, W., Su, Y., Chen, X., Gao, W.: Ensemble of piecewise FDA based on spatial histograms of local (Gabor) binary patterns for face recognition. In: Proc. International Conference on Pattern Recognition, vol. 4, pp. 606–609 (2006)
64. Smith, R.S., Windeatt, T.: Facial expression detection using filtered local binary pattern features with ECOC classifiers and platt scaling. In: JMLR Workshop on Applications of Pattern Analysis, vol. 11, pp. 111–118 (2010)
65. Sokal, R.R., Rohlf, F.J.: Biometry. Freeman, New York (1969)
66. Sun, N., Zheng, W., Sun, C., Zou, C., Zhao, L.: Gender classification based on boosting local binary pattern. In: Proc. International Symposium on Neural Networks, pp. 194–201 (2006)
67. Swain, M.J., Ballard, D.H.: Color indexing. Int. J. Comput. Vis. **7**(1), 11–32 (1991)
68. Tan, N., Huang, L., Liu, C.: A new probabilistic local binary pattern for face verification. In: Proc. IEEE International Conference on Image Processing, pp. 1237–1240 (2009)
69. Tan, X., Triggs, B.: Enhanced local texture feature sets for face recognition under difficult lighting conditions. In: Analysis and Modeling of Faces and Gestures. Lecture Notes in Computer Science, vol. 4778, pp. 168–182. Springer, Berlin (2007)
70. Tan, X., Triggs, B.: Fusing Gabor and LBP feature sets for kernel-based face recognition. In: Analysis and Modeling of Faces and Gestures. Lecture Notes in Computer Science, vol. 4778, pp. 235–249. Springer, Berlin (2007)
71. Trefny, J., Matas, J.: Extended set of local binary patterns for rapid object detection. In: Proc. Computer Vision Winter Workshop, pp. 1–7 (2010)
72. Turtinen, M., Pietikäinen, M.: Contextual analysis of textured scene images. In: Proc. British Machine Vision Conference, pp. 849–858 (2006)
73. Varma, M., Zisserman, A.: A statistical approach to materials classification using image patch exemplars. IEEE Trans. Pattern Anal. Mach. Intell. **31**, 2032–2047 (2009)
74. Wang, L., He, D.C.: Texture classification using texture spectrum. Pattern Recognit. **23**, 905–910 (1990)
75. Wang, X., Han, T.X., Yan, S.: An HOG-LBP human detector with partial occlusion handling. In: Proc. International Conference on Computer Vision, pp. 32–39 (2009)
76. Wolf, L., Hassner, T., Taigman, Y.: Descriptor based methods in the wild. In: Proc. ECCV Workshop on Faces in Real-Life Images, pp. 1–14 (2008)
77. Yan, S., Shan, S., Chen, X., Gao, W.: Locally assembled binary (LAB) feature with feature-centric cascade for fast and accurate face detection. In: Proc. IEEE Conference on Computer Vision and Pattern Recognition, pp. 1–7 (2008)

78. Yang, H., Wang, Y.: A LBP-based face recognition method with Hamming distance constraint. In: Proc. International Conference on Image and Graphics, pp. 645–649 (2007)
79. Yao, C.H., Chen, S.Y.: Retrieval of translated, rotated and scaled color textures. Pattern Recognit. **36**(4), 913–929 (2003)
80. Yu, L., Liu, H.: Feature selection for high-dimensional data: A fast correlation-based filter solution. In: Proc. 12th Int. Conf. on Machine Learning, pp. 856–863 (2003)
81. Zhang, B., Gao, Y., Zhao, S., Liu, J.: Local derivative pattern versus local binary pattern: Face recognition with high-order local pattern descriptor. IEEE Trans. Image Process. **19**(2), 533–544 (2010)
82. Zhang, G., Huang, X., Li, S., Wang, Y., Wu, X.: Boosting local binary pattern (LBP)-based face recognition. In: Proc. Advances in Biometric Person Authentication, pp. 179–186 (2005)
83. Zhang, L., Chu, R.F., Xiang, S.M., Liao, S.C., Li, S.Z.: Face detection based on multi-block LBP representation. In: Proc. IEEE International Conference on Biometrics, pp. 11–18 (2007)
84. Zhang, L., Zhang, L., Guo, Z., Zhang, D.: Monogenic-LBP: A new approach for rotation invariant texture classification. In: Proc. International Conference on Image Processing, pp. 2677–2680 (2010)
85. Zhang, W., Shan, S., Gao, W., Chen, X., Zhang, H.: Local Gabor binary pattern histogram sequence (LGBPHS): A novel non-statistical model for face representation and recognition. In: Proc. International Conference on Computer Vision, vol. 1, pp. 786–791 (2005)
86. Zhao, D., Lin, Z., Tang, X.: Contextual distance for data perception. In: Proc. International Conference on Computer Vision, pp. 1–8 (2007)
87. Zhao, D., Lin, Z., Tang, X.: Laplacian PCA and its applications. In: Proc. International Conference on Computer Vision, pp. 1–8 (2007)
88. Zhao, G., Pietikäinen, M.: Dynamic texture recognition using local binary patterns with an application to facial expressions. IEEE Trans. Pattern Anal. Mach. Intell. **29**(6), 915–928 (2007)
89. Zhao, G., Ahonen, T., Matas, J., Pietikäinen, M.: Rotation invariant image and video description with local binary pattern features. Under review (2011)
90. Zhao, S., Gao, Y., Zhang, B.: Sobel-LBP. In: Proc. International Conference on Image Processing, pp. 2144–2147 (2008)
91. Zhou, H., Wang, R., Wang, C.: A novel extended local-binary-pattern operator for texture analysis. Inf. Sci. **178**, 4314–4325 (2008)
92. Zhu, C., Bichot, C.-E., Chen, L.: Multi-scale color local binary patterns for visual object classes recognition. In: Proc. International Conference on Pattern Recognition, pp. 3065–3068 (2010)

Chapter 3
Spatiotemporal LBP

Dynamic or temporal textures are textures with motion [4, 12]. They encompass the class of video sequences that exhibit some stationary properties in time. The main difference between dynamic texture (DT) and ordinary texture is that the notion of self-similarity, central to conventional image texture, is extended to the spatiotemporal domain [3]. Therefore combining motion and appearance to analyze DT is well justified. Varying lighting conditions greatly affect the gray scale properties of dynamic texture. At the same time, the textures may also be arbitrarily oriented, which suggests using rotation-invariant features. It is important, therefore, to define features which are robust with respect to gray scale changes, rotations and translation. For the analysis of textures in motion, the use of volume local binary patterns and LBP from three orthogonal planes was proposed by Zhao and Pietikäinen [14, 15].

3.1 Basic VLBP

To extend LBP to analysis of image sequences, a neighborhood for an arbitrary point $g_{0,c} = I(x, y, t)$ is defined as follows:

$$g_{i,c} = I(x, y, t + i \times L), \quad i = -1, 0, 1, \tag{3.1}$$

$$g_{i,p} = I(x + x_p, y + y_p, t + i \times L),$$

$$p = 0, \dots, P - 1, i = -1, 0, 1 \quad \text{and} \tag{3.2}$$

$$x_p = x + R\cos(2\pi p/P), \tag{3.3}$$

$$y_p = y - R\sin(2\pi p/P), \tag{3.4}$$

where L is time interval. The local dynamic texture V can thus be defined as the joint distribution v of the gray levels of these $3P + 3$ image pixels. P is the number of local neighboring points around the central pixel in one frame.

$$V = v(g_{-1,c}, g_{-1,0}, \dots, g_{-1,P-1}, g_{0,c}, g_{0,0}, \dots, g_{0,P-1}, g_{1,c}, g_{1,0}, \dots, g_{1,P-1}). \tag{3.5}$$

M. Pietikäinen et al., *Computer Vision Using Local Binary Patterns*,
Computational Imaging and Vision 40,
DOI 10.1007/978-0-85729-748-8_3, © Springer-Verlag London Limited 2011

To get gray-scale invariance, the distribution is thresholded similar to [10]. The gray value of the volume center pixel $(g_{0,c})$ is subtracted from the gray values of the circularly symmetric neighborhood $g_{t,p}$ $(t = -1, 0, 1; p = 0, \cdots, P - 1)$, giving:

$$V = v(g_{-1,c} - g_{0,c}, \; g_{-1,0} - g_{0,c}, \cdots,$$

$$g_{-1,P-1} - g_{0,c}, \; g_{0,c}, \; g_{0,0} - g_{0,c}, \cdots,$$

$$g_{0,P-1} - g_{0,c}, \; g_{1,0} - g_{0,c}, \cdots,$$

$$g_{1,P-1} - g_{0,c}, \; g_{1,c} - g_{0,c}). \tag{3.6}$$

Assuming that differences $g_{t,p} - g_{0,c}$ are independent of $g_{0,c}$ allows the factorization of Eq. 3.6:

$$V \approx v(g_{0,c})v(g_{-1,c} - g_{0,c}, \; g_{-1,0} - g_{0,c}, \cdots,$$

$$g_{-1,P-1} - g_{0,c}, \; g_{0,0} - g_{0,c}, \cdots, g_{0,P-1} - g_{0,c},$$

$$g_{1,0} - g_{0,c}, \cdots, g_{1,P-1} - g_{0,c}, \; g_{1,c} - g_{0,c}).$$

In practice, exact independence is not warranted; hence, the factorized distribution is only an approximation of the joint distribution. However, a possible small loss of information is acceptable as this allows to achieve invariance with respect to shifts in gray scale. Thus, similar to LBP in ordinary texture analysis (Chap. 2), the distribution $v(g_{0,c})$ describes the overall luminance of the image, which is unrelated to the local image texture and, consequently, does not provide useful information for dynamic texture analysis. Hence, much of the information in the original joint gray level distribution (Eq. 3.5) is conveyed by the joint difference distribution:

$$V_1 = v(g_{-1,c} - g_{0,c}, \; g_{-1,0} - g_{0,c}, \cdots,$$

$$g_{-1,P-1} - g_{0,c}, \; g_{0,0} - g_{0,c}, \cdots, g_{0,P-1} - g_{0,c},$$

$$g_{1,0} - g_{0,c}, \cdots, g_{1,P-1} - g_{0,c}, \; g_{1,c} - g_{0,c}).$$

This is a highly discriminative texture operator. It records the occurrences of various patterns in the neighborhood of each pixel in a $(2(P + 1) + P = 3P + 2)$-dimensional histogram.

Invariance with respect to the scaling of the gray scale is achieved by considering simply the signs of the differences instead of their exact values:

$$V_2 = v(s(g_{-1,c} - g_{0,c}), s(g_{-1,0} - g_{0,c}), \cdots,$$

$$s(g_{-1,P-1} - g_{0,c}), s(g_{0,0} - g_{0,c}), \cdots,$$

$$s(g_{0,P-1} - g_{0,c}), s(g_{1,0} - g_{0,c}), \cdots,$$

$$s(g_{1,P-1} - g_{0,c}), s(g_{1,c} - g_{0,c})). \tag{3.7}$$

The expression of V_2 can be simplified as $V_2 = v(v_0, \cdots, v_q, \cdots, v_{3P+1})$, where q corresponds to the index of values in V_2 orderly. By assigning a binomial factor

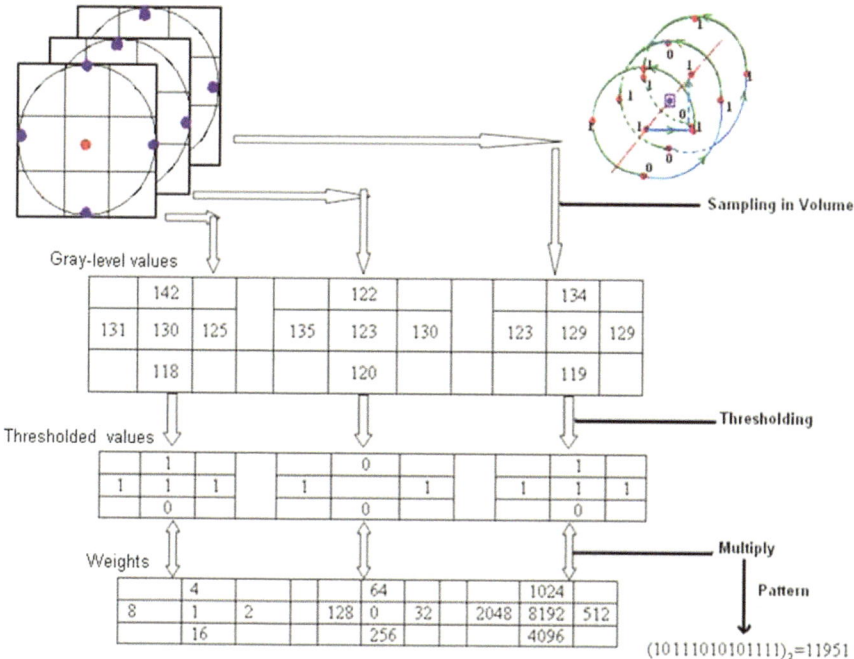

Fig. 3.1 Procedure of VLBP$_{1,4,1}$ [15]

2^q for each sign $s(g_{t,p} - g_{0,c})$, Eq. 3.7 is transformed into a unique VLBP$_{L,P,R}$ number that characterizes the spatial structure of the local volume dynamic texture:

$$\text{VLBP}_{L,P,R} = \sum_{q=0}^{3P+1} v_q 2^q. \tag{3.8}$$

Figure 3.1 shows the whole computing procedure for VLBP$_{1,4,1}$. First, sampling neighboring points in the volume is done. Then, thresholding every point in the neighborhood with the value of the center pixel is performed to get a binary value. Finally, the VLBP code is produced by multiplying the thresholded binary values with weights given to the corresponding pixel and summing up the result.

In calculating VLBP$_{L,P,R}$ distribution for a given $X \times Y \times T$ dynamic texture ($x \in \{0, \cdots, X - 1\}, y \in \{0, \cdots, Y - 1\}, t \in \{0, \cdots, T - 1\}$), the central part is only considered because a sufficiently large neighborhood cannot be used on the borders in this 3D space. The basic VLBP code is calculated for each pixel in the cropped portion of the DT, and the distribution of the codes is used as a feature vector, denoted by D:

$$D = v(\text{VLBP}_{L,P,R}(x, y, t)), \quad x \in \{\lceil R \rceil, \cdots, X - 1 - \lceil R \rceil\},$$
$$y \in \{\lceil R \rceil, \cdots, Y - 1 - \lceil R \rceil\}, t \in \{\lceil L \rceil, \cdots, T - 1 - \lceil L \rceil\}.$$

The histograms are normalized with respect to volume size variations by setting the sum of their bins to unity.

Because the dynamic texture is viewed as sets of volumes and their features are extracted on the basis of those volume textons, VLBP combines the motion and appearance to describe dynamic textures.

3.2 Rotation Invariant VLBP

Dynamic textures may also be arbitrarily oriented, so they also often rotate in the videos. The most important difference between rotation in a still texture image and DT is that the whole sequence of DT rotates around one axis or multi-axes (if the camera rotates during capturing), while the still texture rotates around one point. Therefore, dealing with VLBP as a whole is not possible to get rotation invariant code as in [10], which assumed rotation around the center pixel in the static case. The whole VLBP code from Eq. 3.7 is first divided into 5 parts:

$$
\begin{aligned}
V_2 = v([s(g_{t_c-L,c} - g_{t_c,c})], \\
[s(g_{t_c-L,0} - g_{t_c,c}), \cdots, s(g_{t_c-L,P-1} - g_{t_c,c})], \\
[s(g_{t_c,0} - g_{t_c,c}), \cdots, s(g_{t_c,P-1} - g_{t_c,c})], \\
[s(g_{t_c+L,0} - g_{t_c,c}), \cdots, s(g_{t_c+L,P-1} - g_{t_c,c})], \\
[s(g_{t_c+L,c} - g_{t_c,c})]).
\end{aligned}
$$

This results in $V_{preC}, V_{preN}, V_{curN}, V_{posN}, V_{posC}$ in order, where $V_{preN}, V_{curN}, V_{posN}$ represent the LBP code in the previous, current and posterior frames, respectively, while V_{preC} and V_{posC} represent the binary values of the center pixels in the previous and posterior frames.

$$
\text{LBP}_{t,P,R} = \sum_{p=0}^{P-1} s(g_{t,p} - g_{t_c,c}) 2^p, \quad t = t_c - L, t_c, t_c + L. \qquad (3.9)
$$

Using formula 3.9, $\text{LBP}_{t_c-L,P,R}$, $\text{LBP}_{t_c,P,R}$ and $\text{LBP}_{t_c+L,P,R}$ can be obtained. The effect of rotation is removed by setting:

$$
\begin{aligned}
\text{VLBP}_{L,P,R}^{ri} = \min\{(\text{VLBP}_{L,P,R} \text{ and } 2^{3P+1}) \\
+ \text{ROL}(\text{ROR}(\text{LBP}_{t_c+L,P,R}, i), 2P + 1) \\
+ \text{ROL}(\text{ROR}(\text{LBP}_{t_c,P,R}, i), P + 1) \\
+ \text{ROL}(\text{ROR}(\text{LBP}_{t_c-L,P,R}, i), 1) \\
+ (\text{VLBP}_{L,P,R} \text{ and } 1) | i = 0, 1, \cdots, P - 1\} \qquad (3.10)
\end{aligned}
$$

where $\text{ROR}(x, i)$ performs a circular bit-wise right shift on the P-bit number x i times [10], and $\text{ROL}(y, j)$ performs a bit-wise left shift on the $3P+2$-bit number y

j times. In terms of image pixels, formula 3.10 simply corresponds to rotating the neighbor set in three separate frames clockwise and this happens synchronously so that a minimal value is selected as the VLBP rotation invariant code.

For example, for the original VLBP code $(1, 1010, 1101, 1100, 1)_2$, its codes after rotating clockwise 90, 180, 270 degrees are $(1, 0101, 1110, 0110, 1)_2$, $(1, 1010, 0111, 0011, 1)_2$ and $(1, 0101, 1011, 1001, 1)_2$ respectively. Their rotation invariant code should be $(1, 0101, 1011, 1001, 1)_2$, and not $(00111010110111)_2$ as obtained by using the VLBP as a whole.

In [10], Ojala et al. found that the vast majority of the LBP patterns in a local neighborhood are so called "uniform patterns". A pattern is considered uniform if it contains at most two bitwise transitions from 0 to 1 or vice versa when the bit pattern is considered circular. When using uniform patterns, all non-uniform LBP patterns are stored in a single bin in the histogram computation. This makes the length of the feature vector much shorter and allows to define a simple version of rotation invariant LBP [10]. In the remaining sections, the superscript $riu2$ will be used to denote these features, while the superscript $u2$ means that the uniform patterns without rotation invariance are used. For example, $VLBP_{1,2,1}^{riu2}$ denotes rotation invariant $VLBP_{1,2,1}$ based on uniform patterns.

3.3   Local Binary Patterns from Three Orthogonal Planes

For VLBP, the parameter P determines the number of features. A large P produces a long histogram, while a small P makes the feature vector shorter, but also means losing more information. When the number of neighboring points increases, the number of patterns for basic VLBP will become very large, 2^{3P+2}, as shown in Fig. 3.2. Due to this rapid increase, it is difficult to extend VLBP to have a large number of neighboring points, and this limits its applicability. At the same time, when the time interval $L > 1$, the neighboring frames with a time variance less than L will be omitted.

To address these problems, simplified descriptors are presented by concatenating local binary patterns on three orthogonal planes (LBP-TOP): XY, XT and YT, considering only the co-occurrence statistics in these three directions (shown in Fig. 3.3). Usually a video sequence is thought of as a stack of XY planes in axis T, but a video sequence can also be seen as a stack of XT planes in axis Y and YT planes in axis X, respectively. The XT and YT planes provide information about the space-time transitions. With this approach, the number of bins is only $3 \cdot 2^P$, much smaller than 2^{3P+2}, as shown in Fig. 3.2, which makes the extension to many neighboring points easier and also reduces the computational complexity. There are two main differences between VLBP and LBP-TOP. Firstly, the VLBP uses three parallel planes, of which only the middle one contains the center pixel. The LBP-TOP, on the other hand, uses three orthogonal planes which intersect in the center pixel. Secondly, VLBP considers the co-occurrences of all neighboring points from three parallel frames, which tends to make the feature vector too long. LBP-TOP considers the feature distributions from each separate plane and then concatenates

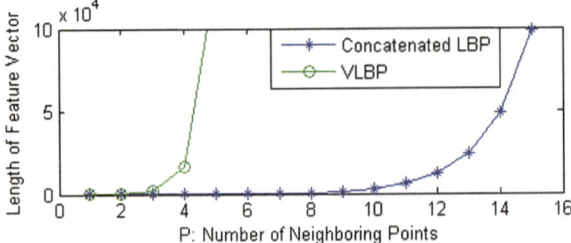

Fig. 3.2 The number of features versus the number of LBP codes. (From G. Zhao and M. Pietikäinen, Dynamic texture recognition using local binary patterns with an application to facial expressions, IEEE Transactions on Pattern Analysis and Machine Intelligence, Vol. 29, Num. 6, 915–928, 2007. @2007 IEEE)

Fig. 3.3 Three planes in DT to extract neighboring points. (From G. Zhao and M. Pietikäinen, Dynamic texture recognition using local binary patterns with an application to facial expressions, IEEE Transactions on Pattern Analysis and Machine Intelligence, Vol. 29, Num. 6, 915–928, 2007. @2007 IEEE)

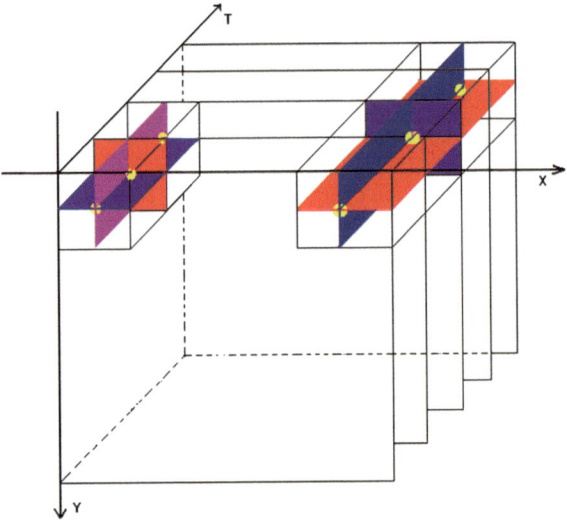

them together, making the feature vector much shorter when the number of neighboring points increases.

To simplify the VLBP for DT analysis and to keep the number of bins reasonable when the number of neighboring points increases, the technique uses three instances of co-occurrence statistics obtained independently from three orthogonal planes [13], as shown in Fig. 3.3. Because the motion direction of textures is unknown, the neighboring points in a circle, and not only in a direct line for central points in time, are considered. Compared with VLBP, not all the volume information, but only the features from three planes are applied. Figure 3.4 demonstrates example images from three planes. (a) shows the image in the XY plane, (b) in the XT plane which gave the visual impression of one row changing in time, while (c) describes the motion of one column in temporal space. The LBP code is extracted from the XY, XT and YT planes, which are denoted as XY-LBP, XT-LBP and YT-LBP, for all pixels, and statistics of three different planes are obtained, and then

Fig. 3.4 (**a**) Image in XY plane (400×300). (**b**) Image in XT plane (400×250) in $y = 120$ (last row is pixels of $y = 120$ in first image). (**c**) Image in TY plane (250×300) in $x = 120$ (first column is the pixels of $x = 120$ in first frame). (From G. Zhao and M. Pietikäinen, Dynamic texture recognition using local binary patterns with an application to facial expressions, IEEE Transactions on Pattern Analysis and Machine Intelligence, Vol. 29, Num. 6, 915–928, 2007. @2007 IEEE)

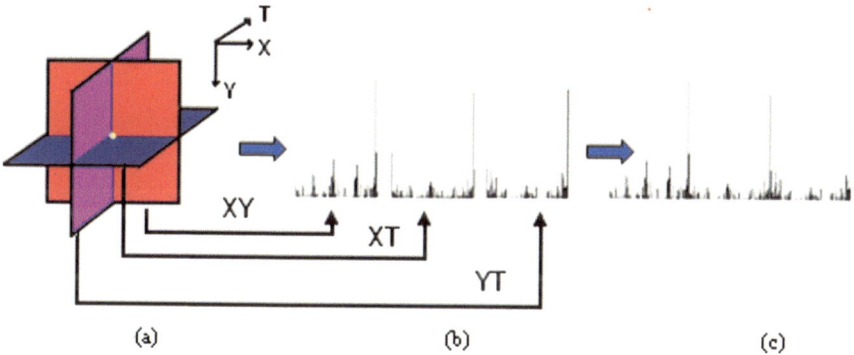

Fig. 3.5 (**a**) Three planes in dynamic texture. (**b**) LBP histogram from each plane. (**c**) Concatenated feature histogram. (From G. Zhao and M. Pietikäinen, Dynamic texture recognition using local binary patterns with an application to facial expressions, IEEE Transactions on Pattern Analysis and Machine Intelligence, Vol. 29, Num. 6, 915–928, 2007. @2007 IEEE)

concatenated into a single histogram. The procedure is demonstrated in Fig. 3.5. In such a representation, DT is encoded by the XY-LBP, XT-LBP and YT-LBP, while the appearance and motion in three directions of DT are considered, incorporating spatial domain information (XY-LBP) and two spatial temporal co-occurrence statistics (XT-LBP and YT-LBP).

Setting the radius in the time axis to be equal to the radius in the space axis is not reasonable for dynamic textures. For instance, for a DT with an image resolution of over 300 by 300, and a frame rate of less than 12, in a neighboring area with a radius of 8 pixels in the X axis and Y axis the texture might still keep its appearance; however, within the same temporal intervals in the T axis, the texture changes drastically, especially in those DTs with high image resolution and a low frame rate. So there are different radius parameters in space and time to set. In the XT and YT planes, different radii can be assigned to sample neighboring points in space and

Fig. 3.6 Different radii and
number of neighboring points
on three planes. (From
G. Zhao and M. Pietikäinen,
Dynamic texture recognition
using local binary patterns
with an application to facial
expressions, IEEE
Transactions on Pattern
Analysis and Machine
Intelligence, Vol. 29, Num. 6,
915–928, 2007. @2007
IEEE)

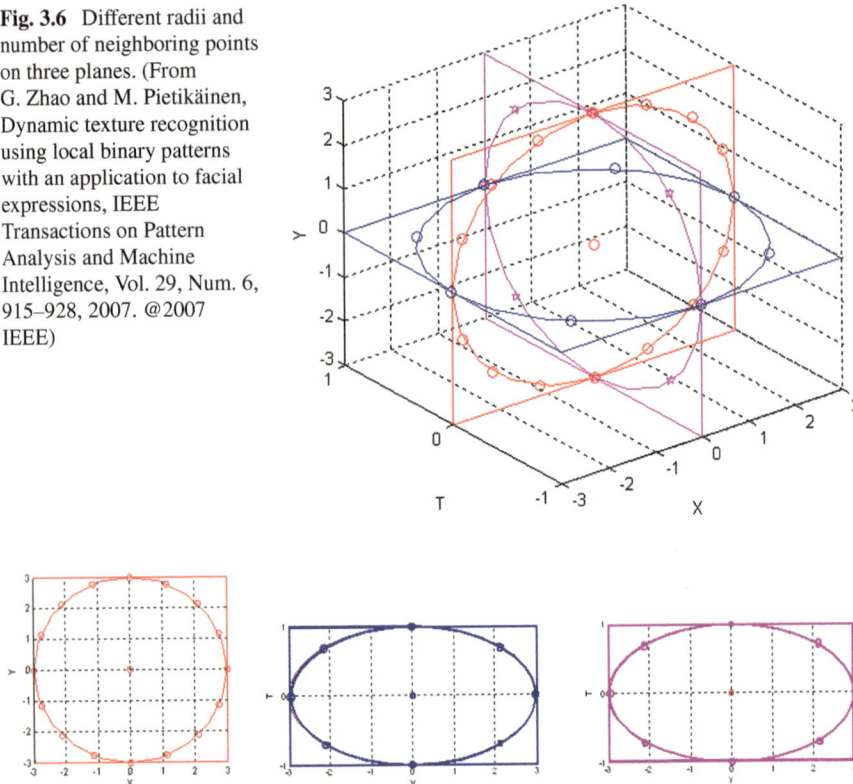

Fig. 3.7 Detailed sampling for Fig. 3.6 with $R_X = R_Y = 3$, $R_T = 1$, $P_{XY} = 16$, $P_{XT} = P_{YT} = 8$.
(**a**) XY plane; (**b**) XT plane; (**c**) YT plane. (From G. Zhao and M. Pietikäinen, Dynamic texture
recognition using local binary patterns with an application to facial expressions, IEEE Transactions
on Pattern Analysis and Machine Intelligence, Vol. 29, Num. 6, 915–928, 2007. @2007 IEEE)

time. With this approach the traditional circular sampling is extended to elliptical
sampling.

More generally, the radii in axes X, Y and T, and the number of neighboring
points in the XY, XT and YT planes can also be different, which can be marked
as R_X, R_Y and R_T, P_{XY}, P_{XT} and P_{YT}, as shown in Fig. 3.6 and Fig. 3.7. The
corresponding feature is denoted as LBP-TOP$_{P_{XY},P_{XT},P_{YT},R_X,R_Y,R_T}$. Suppose the
coordinates of the center pixel $g_{t_c,c}$ are (x_c, y_c, t_c), the coordinates of $g_{XY,p}$ are
given by $(x_c - R_X \sin(2\pi p/P_{XY}), y_c + R_Y \cos(2\pi p/P_{XY}), t_c)$, the coordinates of
$g_{XT,p}$ are given by $(x_c - R_X \sin(2\pi p/P_{XT}), y_c, t_c - R_T \cos(2\pi p/P_{XT}))$, and the
coordinates of $g_{YT,p}$ $(x_c, y_c - R_Y \cos(2\pi p/P_{YT}), t_c - R_T \sin(2\pi p/P_{YT}))$. This is
different from the basic LBP introduced in Chap. 2, and it extends the definition of
LBP.

In calculating LBP-TOP$_{P_{XY},P_{XT},P_{YT},R_X,R_Y,R_T}$ distribution for a given $X \times Y \times T$
dynamic texture ($x_c \in \{0, \cdots, X-1\}$, $y_c \in \{0, \cdots, Y-1\}$, $t_c \in \{0, \cdots, T-1\}$), the

central part is only considered because a sufficiently large neighborhood cannot be
used on the borders in this 3D space.

A histogram of the DT can be defined as

$$H_{i,j} = \sum_{x,y,t} I\left\{f_j(x,y,t) = i\right\}, \quad i = 0, \cdots, n_j - 1; \; j = 0, 1, 2 \qquad (3.11)$$

in which n_j is the number of different labels produced by the LBP operator in the
jth plane ($j = 0$: XY, 1: XT and 2: YT), $f_i(x, y, t)$ expresses the LBP code of
central pixel (x, y, t) in the jth plane, and

$$I\{A\} = \begin{cases} 1, & \text{if } A \text{ is true;} \\ 0, & \text{if } A \text{ is false.} \end{cases}$$

When the DTs to be compared are of different spatial and temporal sizes, the
histograms must be normalized to get a coherent description:

$$N_{i,j} = \frac{H_{i,j}}{\sum_{k=0}^{n_j-1} H_{k,j}}. \qquad (3.12)$$

In this histogram, a description of DT is effectively obtained based on LBP from
three different planes. The labels from the XY plane contain information about the
appearance, and in the labels from the XT and YT planes co-occurrence statistics
of motion in horizontal and vertical directions are included. These three histograms
are concatenated to build a global description of DT with the spatial and temporal
features.

Different number of neighboring points $\{P_{XY}, P_{XT}, P_{YT}\}$ and radii $\{R_X, R_Y, R_T\}$
can be used when computing the features in XY, XT and YT slices to yield the
multiscale features in space and time dimensions [16].

3.4 Rotation Invariant LBP-TOP

3.4.1 Problem Description

A description of DT is effectively obtained based on LBP from three different
planes. However, the appearance-motion planes XT and YT in LBP-TOP are not
rotation invariant, which makes LBP-TOP hard to handle the rotation variations.
This needs to be addressed for DT description. As shown in Fig. 3.8 (right), the
input video in top row is with 60 degrees rotation from that in Fig. 3.8 (left), so
the XY, XT and YT planes in middle row are different from that of Fig. 3.8 (left),
which obviously makes the computed LBP codes different from each other.

Dynamic textures in video sequences can be arbitrarily oriented. The rotation
can be caused by the rotation of cameras and the self-rotation of the captured objects. The most important difference between still images and video sequences is

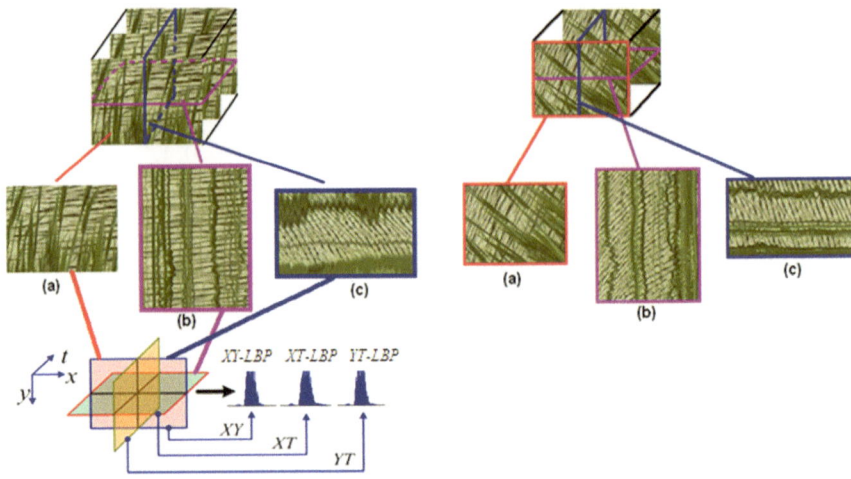

Fig. 3.8 Computation of LBP-TOP for "watergrass" with 0 (*left*) and 60 (*right*) degrees rotation

that in videos the whole sequence rotates around an axis or multi-axes parallel to T axis, while the still image textures rotate around one point. The LBP Histogram Fourier features (LBP-HF) [1] which was introduced in Chap. 2, is extended to an Appearance-Motion (AM) description. Both LBP-XT and LBP-YT describe the appearance and motion. When a video sequence rotates, these two planes do not rotate accordingly, which makes the LBP-TOP operator not rotation invariant. Moreover, rotation only happens around the axis parallel to T axis, so considering the rotation invariant descriptor inside planes does not make any sense. Instead, the rotations of the planes should be considered, not only for the orthogonal planes (XT and YT rotation with 90 degrees), but also for the planes with different rotation angles, like the purple planes with rotations 45 and 135 degrees in Fig. 3.9. So the AM planes consist of P_{XY} rotation planes around T axis. The radius in X, Y and T can be different. Only two types for the number of neighboring points are included, one is P_{XY} which determines how many rotated planes will be considered, the other one is P_T which is the number of neighboring points in AM planes. The original XT and YT are not two separate planes any more, instead they are AM planes obtained by rotating the same plane zero and 90 degrees, respectively.

The corresponding feature is denoted as LBP-TOP$^{ri}_{P_{XY},P_T,R_X,R_Y,R_T}$. Suppose the coordinates of the center pixel $g_{t_c,c}$ are (x_c, y_c, t_c), LBP is computed from P_{XY} spatiotemporal planes. The coordinates of the neighboring points $g_{d,p}$ sampled from the ellipse in XYT space with $g_{t_c,c}$ as center, R_X, R_Y and R_T as the length of axes, are given by $(x_c + R_X \cos(2\pi d/P_{XY})\cos(2\pi p/P_T), y_c - R_Y \sin(2\pi d/P_{XY})\cos(2\pi p/P_T), t_c + R_T \sin(2\pi p/P_T))$, where d $(d = 0, \cdots, P_{XY} - 1)$ is the index of the AM plane and p $(p = 0, \cdots, P_T - 1)$ represents the label of neighboring point in plane d.

Considering "uniform patterns" [10] are usually used to compress the feature vector of LBP and these patterns can be used to represent some texture primitives

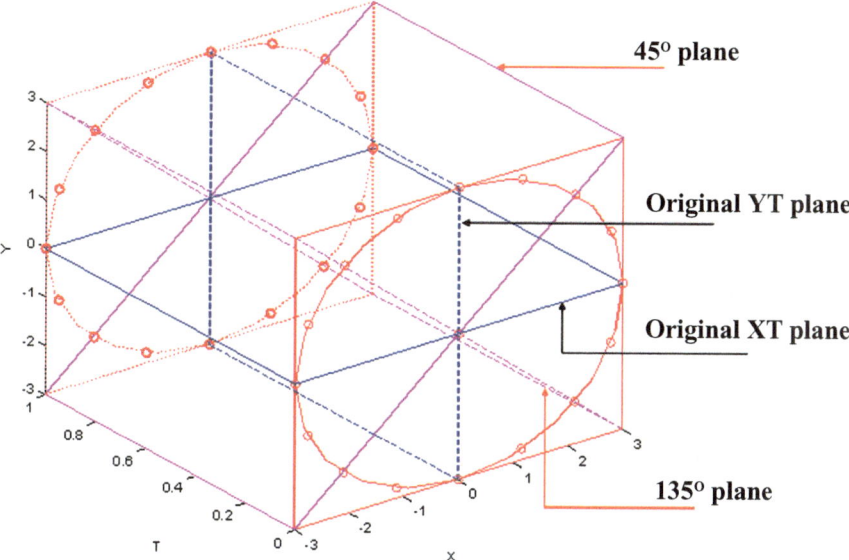

Fig. 3.9 Rotated planes from which LBP is computed

like lines, spots and corners, uniform patterns are used to form new rotation invariant descriptors.

3.4.2 One Dimensional Histogram Fourier LBP-TOP (1DHFLBP-TOP)

After extracting the uniform LBP for all the rotated planes, the Fourier transform for every uniform pattern is computed along all the rotated planes. Figure 3.10 demonstrates the computation. For $P_T = 8$, 59 uniform patterns can be obtained as shown in the left column. For all the rotated $P_{XY} = 8$ planes, discrete Fourier transform is applied for every pattern along all planes to produce the frequency features. Eq. 3.13 illustrates the computation.

$$H_1(n, u) = \sum_{d=0}^{P_{XY}-1} h_1(d, n) e^{-i2\pi u d / P_{XY}}, \qquad (3.13)$$

where n is index of uniform patterns ($n = 0, \cdots, N$, $N = 58$ for $P_T = 8$) and u ($u = 0, \cdots, P_{XY} - 1$) is frequency. d ($d = 0, \cdots, P_{XY} - 1$) is the index of rotation degrees around the line passing through the current central pixel $g_{t_c,c}$ and parallel to T axis. $h_1(d, n)$ is the value of pattern n in uniform LBP histogram at plane d. To get the low frequencies, u can use the value from 0 to ($P_{XY}/s + 1$) (e.g. $s = 4$ in experiments shown in Chap. 7).

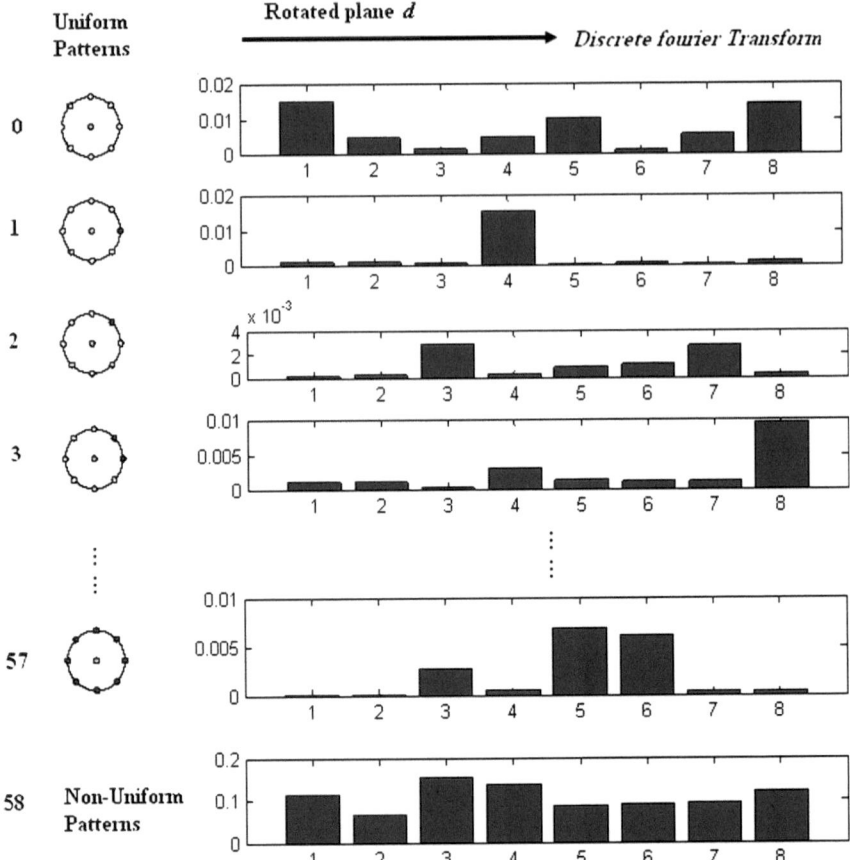

Fig. 3.10 LBP histograms for uniform patterns in different rotated motion planes with $P_{XY} = 8$ and $P_T = 8$

When $u = 0$, $H_1(n, 0)$ means the sum of the pattern n through all the rotated motion planes, which can be thought as another kind of rotation invariant descriptor of simply summing the histograms from all the rotated planes. Since it uses one dimensional histogram Fourier transform for LBP-TOP, it hence can be called *one dimensional histogram Fourier LBP-TOP* (1DHFLBP-TOP).

The feature vector $V1$ of 1DHFLBP-TOP is of the following form:

$$V1 = [\,|H_1(0, 0)|, \cdots, |H_1(0, P_{XY}/s + 1)|, \cdots,$$

$$|H_1(N - 1, 0)|, \cdots, |H_1(N - 1, P_{XY}/s + 1)|\,].$$

N is the number of uniform patterns with neighboring points P_{XY}. Here s is the segments of frequencies. Not all the P_{XY} frequencies are used. Instead only the low frequency, saying $[0\ P_{XY}/s + 1]$ are utilized. Total length of $V1$ is $LRI1 = N \times (P_{XY}/s + 2)$.

One can notice that for the descriptor 1DHFLBP-TOP, the LBPs from a plane rotated g degrees $(180 > g \geq 0)$ and $g + 180$ are mirrored along the T-axis through the central point, but they are not same. So to get the rotation invariant descriptor, all the rotated planes should be used, which increases the computational load. To address this problem, another rotation invariant descriptor for LBP-TOP with 2D Fourier transform: 2DHFLBP-TOP has been proposed. Dynamic texture experiments for view variations can be seen in Chap. 7. More details can be found from [18].

3.5   Other Variants of Spatiotemporal LBP

High success of spatiotemporal LBP methods in various computer vision problems and applications has led to many other teams around the world investigating the approach and several extensions and modifications of Spatiotemporal LBP have been proposed to increase its robustness and discriminative power.

LTP-TOP was developed by Nanni et al. [9]. The encoding function was modified for considering both the ternary patterns and the Three Orthogonal Planes. Their experiments on 10-class Weizmann dataset obtained very good results. Weber Law Descriptor (WLD) has also been extended to spatiotemporal domain in the same way as LBP-TOP, yielding WLD-TOP for supplementing LBP-TOP in dynamic texture segmentation [2].

Mattivi and Shao proposed Extended Gradient LBP-TOP for action recognition [8]. Two modifications were made on the basis of LBP-TOP. Firstly, the computation of LBP was extended to nine slices, three for each axis. Therefore, on the XY dimension there is the original XY plane (centered in the middle of the cuboid) plus other two XY planes located at $1/4$ and $3/4$ of the cuboid's length. The same is done for XT and YT dimensions. Secondly, computation of LBP operator on gradient images was introduced. The gradient image contains information about the rapidity of pixel intensity changes along a specific direction, has large magnitude values at edges and it can further increment LBP operator's performances, since LBP encodes local primitives such as curved edges, spots, flat areas etc. For each cuboid, the brightness gradient is calculated along x, y and t directions, and the resulting three cuboids containing specific gradient information are summed in absolute values. Before computing the image gradients, the cuboid is slightly smoothed with a Gaussian filter in order to reduce noise. The extended LBP-TOP is then performed on the gradient cuboid. Experiments on KTH human action dataset showed the effectiveness of the method.

For dealing with the face recognition, Lei et al. proposed effective LBP operator on three orthogonal planes of Gabor volume (E-GV-LBP) [7]. First, the Gabor face images are formulated as a 3rd-order Gabor volume. Then LBP operator is applied on three orthogonal planes of Gabor volume respectively, named GV-LBP-TOP in short. In this way, the neighboring changes both in spatial space and during different types of Gabor faces can be encoded. Moreover, in order to reduce the computational complexity, an effective GVLBP (E-GV-LBP) descriptor was developed that describes the neighboring changes according to the central point in spatial, scale and

Fig. 3.11 Formulation of
E-GV-LBP. (From Z. Lei, S.
Liao, M. Pietikäinen and S.Z.
Li, Face recognition by
exploring information jointly
in space, scale and
orientation, IEEE
Transactions on Image
Processing, Vol. 20, 247–256,
2011. @2011 IEEE)

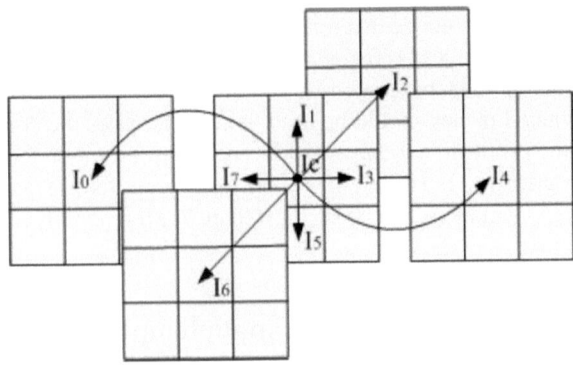

orientation domains simultaneously for face representation. Figure 3.11 shows the formulation of E-GV-LBP.

Visual information from captured video is important for speaker identification under noisy conditions. Combination of LBP dynamic texture and EdgeMap structural features were proposed to take both motion and appearance into account [17], providing the description ability for spatiotemporal development in speech. Spatiotemporal dynamic texture features of local binary patterns extracted from localized mouth regions are used for describing motion information in utterances, which can capture the spatial and temporal transition characteristics. Structural edge map features are extracted from the image frames for representing appearance characteristics. Combination of dynamic texture and structural features takes both motion and appearance together into account, providing the description ability for spatiotemporal development in speech (Fig. 3.12). In the experiments on BANCA and XM2VTS databases, the proposed method obtained promising recognition results comparing to the other features.

Goswami et al. proposed a novel approach to ordinal contrast measurement called Local Ordinal Contrast Patterns(LOCP) [6]. Instead of computing the ordinal contrast with respect to any fixed value such as that at the center pixel or the average intensity value, it computes the pairwise ordinal contrasts for the chain of pixels representing the circular neighborhoods starting from the center pixel. Then it was extended for dynamic texture analysis by extracting the LOCP in three orthonormal planes to generate LOCP-TOP. Together with LDA, its performance of mouthregion biometrics in the XM2VTS database received good results.

LBP3D was proposed by Paulhac et al. for characterization of three-dimensional textures [11]. In their three-dimensional extension of local binary pattern method, the neighbors were defined in a sphere. For a central voxel g_c with the coordinates $(0, 0, 0)$, the coordinates of $g_{pp'}$ are given by $(R \cos((\pi p')/(S - 1)) \cos((2\pi p)/P)$, $R \cos((\pi p')/(S - 1)) \sin((2\pi p)/P), R \sin((p'\pi)/(S - 1)))$, where R is a sphere radius, S the number of circle used to represent the sphere, and P the number of vertex in each circle (Fig. 3.13). The 3D LBP texture operator can then be defined

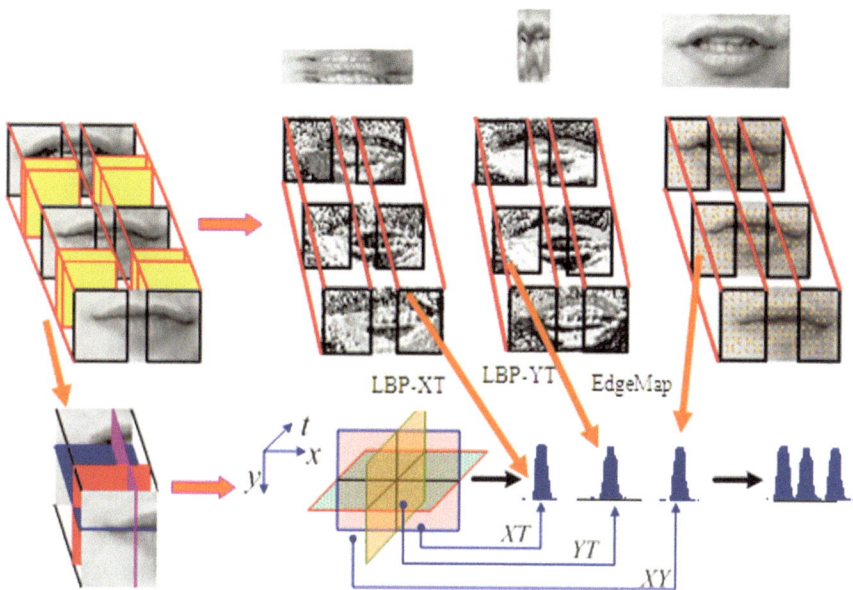

Fig. 3.12 Local spatiotemporal feature extraction

as follows:

$$\mathrm{LBP}_{P',R}^{riu2} = \begin{cases} \sum_{p=0}^{P'-1} s(g_p - g_c) & \text{if } U(\mathrm{LBP}_{P',R}) \leq V) \\ P' + 1 & \text{otherwise} \end{cases} \tag{3.14}$$

with

$$P' = (S-2) \times P + 2, \qquad s(x) = \begin{cases} 1, & x \geq 0; \\ 0, & x < 0. \end{cases}$$

As in two-dimensional case, U is an uniformity measure function that counts the number of uniform black and white regions in the three-dimensional LBP pattern. To allow this operation, a graph is first constructed using all the points on the sphere. Each vertex of the sphere is connected with its closest neighbors to obtain a related graph. Using this graph, a region growing algorithm is applied to identify regions in the three-dimensional pattern. In two dimensions, a pattern is defined as uniform when the number of regions is lower than two. In three dimensions, this condition is relaxed with $V \in \{2, 3\}$. A drawback is that this descriptor is not rotation invariant.

Extension of the original LBP from 2D images to 3D volume data was also considered by Fehr and Burkhardt, achieving a full rotation invariance in 3D [5]. Firstly, a 2-pattern P_N^r is the volume representation (3D grid) of a set of N equidistant points on a sphere with radius r. Each of these points is weighted in an arbitrary but fixed order with the gray values $p_0 := 2^0, \cdots, p_{N-1} = 2^{N-1}$. All other points in the volume are set to zero. Given a center point with gray value c, the point-wise threshold of the entire volume grid is computed to get the thresholded vector T^r. Then the

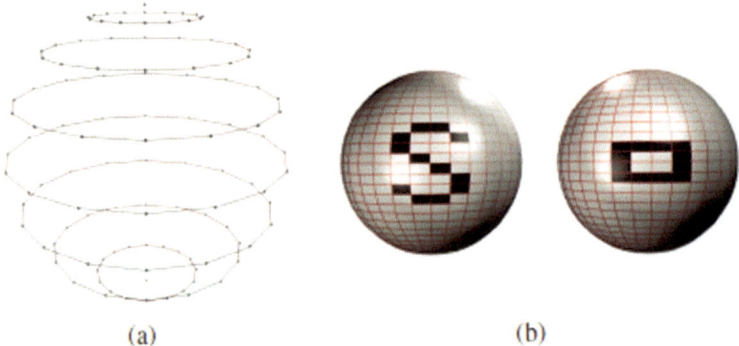

(a)   (b)

Fig. 3.13 (**a**) Representation of a 3D local binary pattern ($S = 9$, $R = 2$, $P = 16$). (**b**) Example of two 3D Local Binary Pattern with the same LBP code (LBP$^{riu2}_{P',R}$ = 12). Here, the value of black vertex is 1 and the value of the others is 0 [11]

rotation invariance is formulated as the computation of the minimum of the full correlation ($\star$) over all angles of the fixed 2-pattern P^r_N with T^r. The methods were evaluated in the context of 3D texture analysis of biological data.

References

1. Ahonen, T., Matas, J., He, C., Pietikäinen, M.: Rotation invariant image description with local binary pattern histogram Fourier features. In: Scandinavian Conference on Image Analysis. Lecture Notes in Computer Science, vol. 5575, pp. 61–70. Springer, Berlin (2009)
2. Chen, J., Zhao, G., Pietikäinen, M.: An improved local descriptor and threshold learning for unsupervised dynamic texture segmentation. In: Proc. ICCV Workshop on Machine Learning for Vision-Based Motion Analysis, pp. 460–467 (2009)
3. Chetverikov, D., Péteri, R.: A brief survey of dynamic texture description and recognition. In: Proc. International Conference on Computer Recognition Systems, pp. 17–26 (2005)
4. Doretto, G., Chiuso, A., Soatto, S., Wu, Y.N.: Dynamic textures. Int. J. Comput. Vis. **51**, 91–109 (2003)
5. Fehr, J., Burkhardt, H.: 3D rotation invariant local binary patterns. In: Proc. International Conference on Pattern Recognition, pp. 1–4 (2008)
6. Goswami, B., Chan, C.H., Kittler, J., Christmas, B.: Local ordinal contrast pattern histograms for spatiotemporal, lip-based speaker authentication. In: Fourth IEEE International Conference on Biometrics: Theory Applications and Systems (BTAS), pp. 1–6 (2010)
7. Lei, Z., Liao, S., Pietikäinen, M., Li, S.Z.: Face recognition by exploring information jointly in space, scale and orientation. IEEE Trans. Image Process. **20**, 247–256 (2011)
8. Mattivi, R., Shao, L.: Human action recognition using LBP-TOP as sparse spatio-temporal feature descriptor. In: Proc. International Conference on Computer Analysis of Images and Patterns, pp. 740–747 (2009)
9. Nanni, L., Brahnam, S., Lumini, A.: Local ternary patterns from three orthogonal planes for human action classification. Expert Syst. Appl. **38**, 5125–5128 (2011)
10. Ojala, T., Pietikäinen, M., Mäenpää, T.: Multiresolution gray-scale and rotation invariant texture classification with local binary patterns. IEEE Trans. Pattern Anal. Mach. Intell. **24**(7), 971–987 (2002)

11. Paulhac, L., Makris, P., Ramel, J.Y.: Comparison between 2D and 3D local binary pattern methods for characterization of three-dimensional textures. In: Proc. International Conference on Image Analysis and Recognition, pp. 670–679 (2008)
12. Szummer, M., Picard, R.W.: Temporal texture modeling. In: Proc. IEEE International Conference on Image Processing, vol. 3, pp. 823–826 (1996)
13. Zhao, G., Pietikäinen, M.: Local binary pattern descriptors for dynamic texture recognition. In: Proc. International Conference on Pattern Recognition, vol. 2, pp. 211–214 (2006)
14. Zhao, G., Pietikäinen, M.: Dynamic texture recognition using local binary patterns with an application to facial expressions. IEEE Trans. Pattern Anal. Mach. Intell. 29(6), 915–928 (2007)
15. Zhao, G., Pietikäinen, M.: Dynamic texture recognition using volume local binary patterns. In: Dynamical Vision. Lecture Notes in Computer Science, vol. 4358, pp. 165–177. Springer, Berlin (2007)
16. Zhao, G., Pietikäinen, M.: Boosted multi-resolution spatiotemporal descriptors for facial expression recognition. Pattern Recognit. Lett. 30(12), 1117–1127 (2009)
17. Zhao, G., Huang, X., Gizatdinova, Y., Pietikäinen, M.: Combining dynamic texture and structural features for speaker identification. In: Proc. ACM Multimedia Workshop on Multimedia in Forensics, Security and Intelligence, pp. 93–98 (2010)
18. Zhao, G., Ahonen, T., Matas, J., Pietikäinen, M.: Rotation invariant image and video description with local binary pattern features. Under review (2011)

Part II
Analysis of Still Images

Chapter 4
Texture Classification and Segmentation

In texture classification, the aim is to assign an unseen texture sample into one of predefined classes. The assignment is done based on rules which are typically derived automatically from a training set consisting of texture samples with known classes. The LBP methodology has performed very well in many comparative studies performed using publicly available texture image datasets. The first part of this chapter gives an introduction to some datasets and comparative studies.

Image segmentation is an important step in most applications of computer vision. Since the work on LBP-based unsupervised texture segmentation in the late 1990s [14, 15], the LBP operator has become very popular in texture-based image segmentation. The second part of this chapter introduces the original method using distributions of LBP/C features for measuring the similarity of adjacent regions during the segmentation process. A region-based algorithm is used for coarse image segmentation and a pixelwise classification scheme for improving localization of region boundaries.

4.1 Texture Classification

Given a segmented texture image to be classified, the two critical components are the feature extractor and the classification algorithm. LBP based descriptors were described in the previous chapters, and the reader is referred to [13] for a substantive listing of other texture descriptors presented in the literature. For a general introduction and review of statistical pattern classifiers, refer to [2, 7], for example. In texture classification, the nearest neighbor or k nearest neighbor classifier with different distance measures is a common choice, e.g. [16, 23]. Recently, the Support Vector Machine (SVM) [22] has gained major interest, and it has been reported to outperform nearest neighbor classifier in texture classification in [9] and [26], for example. Also, boosting based approaches such as AdaBoost [8], and bagging classifiers like the Random Forest classifier [3] have been successfully applied to texture classification. The problem of texture retrieval is to some extent related to texture classification. In essence, texture retrieval is content based image retrieval applied

M. Pietikäinen et al., *Computer Vision Using Local Binary Patterns*,
Computational Imaging and Vision 40,
DOI 10.1007/978-0-85729-748-8_4, © Springer-Verlag London Limited 2011

Fig. 4.1 Brodatz textures

to texture images. Thus, the aim is to retrieve from a database as many samples of a requested texture as possible. However, texture is more often used as an additional feature for general image retrieval.

4.1.1 Texture Image Datasets

Classification of texture images in itself is seldom the final goal or actual application of a texture analysis system; texture classification is more often used to evaluate the performance of texture features or a classification algorithm.

There are several texture image sets that are commonly used in the texture analysis literature and that are publicly available for experiments. In the following, the most commonly used texture image sets are described.

The Brodatz texture images [4] is one of the oldest and most widely used texture image sets used in texture classification experiments. Altogether, the Brodatz Album contains 112 images, each representing a different texture, some of which are shown in Fig. 4.1. Originally, the album was not intended for texture classification experiments, and there is no standard test protocol associated with it. Typically,

Fig. 4.2 KTH-TIPS2 textures

some subset of the 112 images has been used in classification experiments. The complete scanned texture image is divided into subimages, part of which is then used for training and another part for testing. Because of lack of a standard protocol, the classification results on Brodatz images from different sources are usually not comparable to each other.

The Columbia-Utrecht Reflectance and Texture database (CUReT) [6] contains images of 61 different real world textures each imaged under more than 200 different viewpoints and under different illuminations (www1.cs.columbia.edu/CAVE/software/curet/). Examples of 61 CUReT textures are shown in Sect. 6.2.3. In total, the database consists of over 14000 images. As with the Brodatz textures, there is no fixed experimental protocol for using the CUReT textures.

Images in the CUReT set contain very little variation in scale. The KTH-TIPS databases [12] were created to extend the CUReT set with images having variation in scale, and with new samples of materials contained by the CUReT set (www.nada.kth.se/cvap/databases/kth-tips/). The KTH-TIPS1 database consists of images of 10 textured materials imaged at 81 different scale, orientation and lighting settings. The KTH-TIPS2 database (Fig. 4.2), on the other hand, is intended for experiments in material classification, and it contains images of different samples of the same material. Eleven different materials are included, and there are four samples of each material. Each sample is imaged under 108 different conditions. There is no strict protocol defined for using the KTH-TIPS1 or KTH-TIPS2 datasets.

The Outex texture image set (www.outex.oulu.fi) [18] is a collection of images of 319 different textures, as shown in Fig. 4.3. Each texture has been imaged under different rotations, scales, and lighting conditions. Furthermore, several differ-

Fig. 4.3 Outex textures

ent experiments with precisely defined training and testing sets and performance measurements have been published. The experiments do not consider only texture classification, but also texture retrieval and supervised and unsupervised texture segmentation.

The UIUC texture image set [10] is one of the newest texture image sets. It contains 25 different textures, and 40 images of each texture. The images have strong variation in viewing angles, and also non-rigid deformations of the textures.

4.1.2 Texture Classification Experiments

This section reviews some of the texture classification experiments reported in the literature which have compared LBP to other texture descriptors.

In [17], Ojala et al. performed extensive tests to validate the rotation invariant LBP operator for texture classification. The experiments included tests with Brodatz and Outex textures. On the Outex_TC_0010 test suite, intended for testing rotation invariant texture classifiers, multi-resolution LBP operator achieved classification accuracy of 96.1% and when combined with local variance, 97.7%. Wavelet based rotation invariant features that were used as a control method, achieved classification accuracy of 80.4%.

Caputo et al. [5] performed experiments in material classification using the KTH-TIPS2 dataset. They used multi-scale LBP features at 3 and 4 different scales as well as VZ-Joint [24] and MR8 [23] features. As classifiers, two nearest neighbor and two different types of support vector machine classifiers were used. In the experiments it was shown that material categorization is a difficult problem and not nearly as good results as in normal texture classification are obtained. In terms of classification accuracy, there was no significant difference between the feature types compared. The choice of classifier was however shown to be an important factor, and SVMs performed clearly better than the nearest neighbor classifier. When using one sample per material for training and 4-scale LBP features, nearest neighbor classifier resulted in categorization accuracy of about 50% and class-specific SVM introduced in the paper gave a categorization rate of about 70%.

Ahonen and Pietikäinen [1] performed an experimental analysis of the significance of different design choices in the LBP operator. Differencing (see Eq. 2.7) was compared to filtering with Gabor or MR8 filters, and thresholding (Eq. 2.9) was compared to codebook based vector quantization. In texture classification, experiments were performed on the KTH-TIPS2 dataset. The material classification experiments showed that in this dataset, thresholding based labeling yields slightly better categorization rates than codebook based. Also, local differencing was shown to result in higher categorization rates than using Gabor or MR8 filters.

4.2 Unsupervised Texture Segmentation

Segmentation of an image into differently textured regions is often a very difficult problem. Usually one does not know a priori what types of textures exist in an image, how many textures there are, and what regions have which textures. In order to distinguish reliably between two textures, relatively large samples of them must be examined, i.e. relatively large blocks of the image. But a large block is unlikely to be entirely contained in a homogeneously textured region and it becomes difficult to correctly determine the boundaries between regions. The performance of texture segmentation is highly dependent on the texture features used. The features should easily discriminate various types of textures. The window size used for computing texture features should be small enough to be useful for small image regions and to provide small error rates at region boundaries. The LBP method meets well these requirements.

An unsupervised texture segmentation algorithm using the LBP/C texture measure and nonparametric statistical test was introduced by Ojala and Pietikäinen [14, 15]. The method performed very well in experiments. It is not sensitive to the selection of parameter values, does not require any prior knowledge about the number of textures or regions in the image as many other approaches do, and provided better results than other approaches. The method can be easily generalized, e.g., to utilize other texture features, multiscale information, color features, and combinations of multiple features.

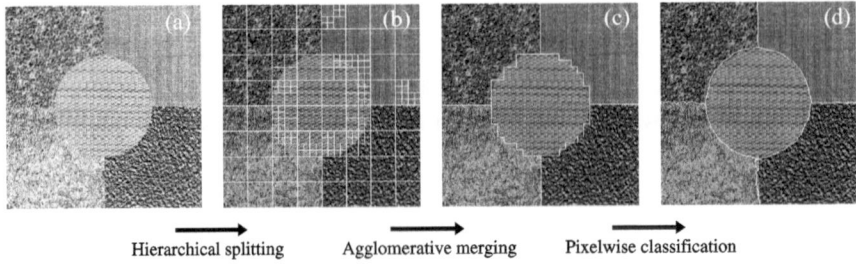

Hierarchical splitting   Agglomerative merging   Pixelwise classification

Fig. 4.4 Main sequence of the segmentation algorithm. (**a**) Original image, (**b**) hierarchical splitting, (**c**) agglomerative merging, (**d**) pixelwise classification

4.2.1 Overview of the Segmentation Algorithm

The segmentation method contains three phases: hierarchical splitting, agglomerative merging and pixelwise classification. First, hierarchical splitting is used to divide the image into regions of roughly uniform texture. Then, an agglomerative merging procedure merges similar adjacent regions until a stopping criterion is met. At this point, rough estimates of the different textured regions present in the image have been obtained, and the analysis in completed by a pixelwise classification to improve the localization. Figure 4.4 illustrates the steps of the segmentation algorithm on a 512×512 mosaic containing five different Brodatz textures [14].

In the original segmentation procedure described below, a log-likelihood-ratio, the G statistic, was used as a pseudo-metric for comparing LBP/C distributions [20]. The value of the G statistic indicates the probability that the two sample distributions come from the same population: the higher the value, the lower the probability that the two samples are from the same population. The similarity of two histograms was measured with a two-way test of interaction or heterogeneity:

$$
G = 2\left(\left[\sum_{s,m}\sum_{i=1}^{n} f_i \log f_i\right] - \left[\sum_{s,m}\left(\sum_{i=1}^{n} f_i\right)\log\left(\sum_{i=1}^{n} f_i\right)\right]\right.
$$
$$
\left. - \left[\sum_{i=1}^{n}\left(\sum_{s,m} f_i\right)\log\left(\sum_{s,m} f_i\right)\right] + \left[\left(\sum_{s,m}\sum_{i=1}^{n} f_i\right)\log\left(\sum_{s,m}\sum_{i=1}^{n} f_i\right)\right]\right), \quad (4.1)
$$

where s, m are the two sample histograms, n is the number of bins and f_i is the frequency at bin i. The more alike the histograms s and m are, the smaller is the value of G.

In later studies adopting this methodology, the G statistic has been often replaced with the simpler histogram intersection algorithm [21].

4.2.2 Splitting

A necessary prerequisite for the agglomerative merging to be successful is that the individual image regions be uniform in texture. For this purpose, the hierarchical splitting algorithm is applied, which recursively splits the original image into square blocks of varying size. The decision on whether a block is split into four subblocks is based on a uniformity test. Using Eq. 4.2 the six pairwise G distances between the LBP/C histograms of the four subblocks are measured. By denoting the largest of the six G values by G_{max}, and the smallest by G_{min}, the block is found to be non-uniform and is thus split further into four subblocks, if a measure of relative dissimilarity within the region is greater than a threshold:

$$R = \frac{G_{max}}{G_{min}} > X. \tag{4.2}$$

Regarding the proper choice of X, one should rather choose too small a value for X than too large a value. It is better to split too much than too little, for the following agglomerative merging procedure is able to correct errors, in cases where a uniform block of a single texture has been needlessly split, but error recovery is not possible if segments containing several textures are assumed to be uniform.

To begin with, the image is divided into rectangular blocks of size S_{max}. If the uniformity test would be applied to arbitrarily large image segments, the detection of small texture patches could fail and end up treating regions containing several textures as uniform.

The next step is to use the uniformity test. If a block does not satisfy the test, it is divided into four subblocks. This procedure is repeated recursively on each subblock until a predetermined minimum block size S_{min} is reached. It is necessary to set a minimum limit on the block size, because the block has to contain a sufficient number of pixels for the LBP/C histogram to be reliable.

Figure 4.4(b) illustrates the result of the hierarchical splitting algorithm with $X = 1.2$, $S_{max} = 64$ and $S_{min} = 16$. As expected, the splitting goes deepest around the texture boundaries.

4.2.3 Agglomerative Merging

Once the image has been split into blocks of roughly uniform texture, an agglomerative merging procedure is applied, which merges similar adjacent regions until a stopping criterion is satisfied. At a particular stage of the merging, that pair of adjacent segments is merged which has the smallest Merger Importance (MI) value. MI is defined as $MI + p \times G$ where p is the number of pixels in the smaller of the two regions and G is the distance measure defined earlier. In other words, at each step the procedure chooses that merge, of all possible merges, which introduces the smallest change in the segmented image. Once the pair of adjacent segments with the smallest MI value has been found, the regions are merged and the two respective LBP/C histograms are summed to be the histogram of the new region. Before moving to the next merge, the G distances between the new region and all regions adjacent to it are computed. Merging is allowed to proceed until the stopping rule

$$\text{MIR} = \frac{\text{MI}_{cur}}{\text{MI}_{max}} > Y \qquad (4.3)$$

triggers. Merging is halted if MIR, the ratio of MI_{cur}, the Merger Importance for the current best merge, and MI_{max}, the largest Merger Importance of all the preceding merges, exceeds a preset threshold Y. Threshold Y determines the scale of texture differences in the segmentation result and therefore the choice of Y depends on the application. In theory, it is possible that the very first merges have a zero MI value (i.e. there are adjacent regions with identical LBP/C histograms), which would lead to a too early termination of the agglomerative merging phase. To prevent this the stopping rule is not evaluated for the first 10% of all possible merges.

Figure 4.4(c) shows the result of the agglomerative merging phase after 174 merges. The MIR of the 175th merge is 9.5 and the merging procedure stops. For comparison, the highest MIR value up to that point had been 1.2.

4.2.4 Pixelwise Classification

If the hierarchical splitting and agglomerative merging phases have been successful, quite reliable estimates of the different textured regions present in the image have been obtained. Treating the LBP/C histograms of the image segments as the texture models, a texture classification mode is started. If an image pixel is on the boundary of at least two distinct textures (i.e. the pixel is 4-connected to at least one pixel with a different label), a discrete disc with radius r is placed on the pixel and the LBP/C histogram over the disc is computed. The G distances between the histogram of the disc and the models of those regions which are 4-connected to the pixel in question are computed. The pixel is relabeled, if the label of the nearest model is different from the current label of the pixel and there is at least one 4-connected adjacent pixel with the tentative new label. The latter condition improves smooth adaption of texture boundaries and decreases the probability of small holes occurring inside the regions. If the pixel is relabeled, i.e. it is moved from an image segment to the adjacent segment, the corresponding texture models are updated accordingly, hence the texture models become more accurate during the process. Only those pixels at which the disc is entirely inside the image are examined, hence the final segmentation result will contain a border of r pixels wide.

In the next scan over the image only the neighborhoods of those pixels are checked, which were relabeled in the previous sweep. The process of pixelwise classification continues until no pixels are relabeled or maximum number of sweeps is reached. This is set to be two times S_{min} based on the reasoning that the boundary estimate of the agglomerative merging phase can be at most this far away from

Fig. 4.5 Segmentation of a natural scene

the "true" texture boundary. Setting an upper limit for the number of iterations ensures that the process will not wander around endlessly, if the disc is not able to capture enough information of the local texture to be stable. According to the experiments the algorithm generally converges quickly with homogeneous textures, whereas with locally stochastic natural scenes maximum number of sweeps may be consumed.

Figure 4.4(d) demonstrates the final segmentation result after the pixelwise classification phase. A disc with a radius of 11 pixels was used and 16 scans were needed. The segmentation error is 1.7%.

4.2.5   Experiments

This segmentation method has been applied to a variety of texture mosaics with very good results. The segmentation of a natural scene is shown in Fig. 4.5 [14]. The textures of natural scenes are generally more non-uniform than the homogeneous textures of the test mosaics. Also, in natural scenes adjacent textured regions are not necessarily separated by well-defined boundaries, but the spatial pattern smoothly changes from one texture to another. Further, the infinite scale of texture differences present in natural scenes can be observed; choosing the right scale is a very subjective matter. For these reasons there is often no 'correct' segmentation for a natural scene, as is the case with texture mosaics.

The invariance of the LBP/C transform to average luminance shows in the bottom part of the image, where the sea is interpreted as a single region despite the shadows. The result obtained is very satisfactory, considering that important color or gray scale information is not utilized in the segmentation.

4.3   Discussion

The LBP method has provided a state-of-the-art performance in many classification experiments with various publicly available texture datasets. Usually a multi-scale version of the LBP should be preferred to achieve higher performance. In such applications in which illumination invariance is not a relevant issue, the LBP combined with a complementary contrast measure usually performs better that LBP alone.

Due to the simplicity of the LBP method and its high performance, the LBP-based segmentation approach has become very popular. It has been adopted and further developed by many researchers. For example, Whelan and Ghita used it as a basis in their very effective color texture segmentation algorithm [25]. In this method the color and texture information are combined adaptively in a composite image descriptor. LBP/C is used as the texture descriptor and color information is obtained by using an Expectation-Maximization space partitioning technique. Color and texture distributions are used as input to a split and merge type algorithm very similar to the presented method. Pixelwise classification at region boundaries is used to provide final segmentation. The method has been successfully applied to the segmentation of natural, medical and industrial images. In other studies, the LBP method has been successfully adopted for the segmentation of remote sensing images, see for example [11]. In Chap. 7, this method will be used as a basis in developing a method for unsupervised segmentation of dynamic textures. Another type of texture segmentation based on LBP-guided active contours has been proposed by Savelonas et al. [19]. Experiments with images of texture mosaics and natural scenes indicate that their method provides comparable or better performance than many existing methods with less computational effort.

References

1. Ahonen, T., Pietikäinen, M.: Image description using joint distribution of filter bank responses. Pattern Recognit. Lett. **30**(4), 368–376 (2009)
2. Bishop, C.M.: Pattern Recognition and Machine Learning. Springer, New York (2006)
3. Breiman, L.: Random forests. Mach. Learn. **45**(1), 5–32 (2001)
4. Brodatz, P.: Textures; A Photographic Album for Artists and Designers. Dover, New York (1966)
5. Caputo, B., Hayman, E., Mallikarjuna, P.: Class-specific material categorisation. In: Proc. International Conference on Computer Vision, pp. 1597–1604 (2005)
6. Dana, K.J., van Ginneken, B., Nayar, S.K., Koenderink, J.J.: Reflectance and texture of real-world surfaces. ACM Trans. Graph. **18**(1), 1–34 (1999)
7. Duda, R.O., Hart, P.E., Stork, D.G.: Pattern Classification. Wiley, New York (2001)
8. Freund, Y., Schapire, R.: A decision-theoretic generalization of on-line learning and an application to boosting. J. Comput. Syst. Sci. **55**(1), 119–139 (1997)
9. Hayman, E., Caputo, B., Fritz, M., Eklundh, J.-O.: On the significance of real-world conditions for material classification. In: 8th European Conference on Computer Vision (ECCV 2004). Lecture Notes in Computer Science, vol. 3024, pp. 253–266. Springer, Berlin (2004)
10. Lazebnik, S., Schmid, C., Ponce, J.: A sparse texture representation using local affine regions. IEEE Trans. Pattern Anal. Mach. Intell. **27**(8), 1265–1278 (2005)
11. Lucieer, A., Stein, A., Fisher, P.: Multivariate texture-based segmentation of remotely sensed imagery for extraction of objects and their uncertainty. Int. J. Remote Sens. **26**, 2917–2936 (2005)
12. Mallikarjuna, P., Fritz, M., Targhi, A.T., Hayman, E., Caputo, B., Eklundh, J.-O.: The KTH-TIPS and KTH-TIPS2 databases, 2006. http://www.nada.kth.se/cvap/databases/kth-tips/

13. Mirmehdi, M., Xie, X., Suri, J. (eds.): Handbook of Texture Analysis. Imperial College Press, London (2008)
14. Ojala, T., Pietikäinen, M.: Unsupervised texture segmentation using feature distributions. In: International Conference on Image Analysis and Processing. Lecture Notes in Computer Science, vol. 1310, pp. 311–318. Springer, Berlin (1997)
15. Ojala, T., Pietikäinen, M.: Unsupervised texture segmentation using feature distributions. Pattern Recognit. **32**, 477–486 (1999)
16. Ojala, T., Pietikäinen, M., Harwood, D.: A comparative study of texture measures with classification based on feature distributions. Pattern Recognit. **29**(1), 51–59 (1996)
17. Ojala, T., Pietikäinen, M., Mäenpää, T.: Multiresolution gray-scale and rotation invariant texture classification with local binary patterns. IEEE Trans. Pattern Anal. Mach. Intell. **24**(7), 971–987 (2002)
18. Ojala, T., Mäenpää, T., Pietikäinen, M., Viertola, J., Kyllönen, J., Huovinen, S.: Outex—new framework for empirical evaluation of texture analysis algorithms. In: Proc. International Conference on Pattern Recognition, pp. 701–706 (2002)
19. Savelonas, M.A., Iakovidis, D.K., Maroulis, D.: LBP-guided active contours. Pattern Recognit. Lett. **29**(9), 1404–1415 (2008)
20. Sokal, R.R., Rohlf, F.J.: Biometry. Freeman, New York (1969)
21. Swain, M.J., Ballard, D.H.: Color indexing. Int. J. Comput. Vis. **7**(1), 11–32 (1991)
22. Vapnik, V. (ed.): Statistical Learning Theory. Wiley, New York (1998)
23. Varma, M., Zisserman, A.: Classifying images of materials: Achieving viewpoint and illumination independence. In: European Conference on Computer Vision. Lecture Notes in Computer Science, vol. 2352, pp. 255–271. Springer, Berlin (2002)
24. Varma, M., Zisserman, A.: Texture classification: Are filter banks necessary? In: Proc. IEEE Conference on Computer Vision and Pattern Recognition, vol. 2, pp. 691–698 (2003)
25. Whelan, P.F., Ghita, O.: Colour texture analysis. In: Mirmehdi, M., Xie, X., Suri, J. (eds.) Handbook of texture analysis, pp. 129–163. Imperial College Press, London (2008)
26. Zhang, J., Marszalek, M., Lazebnik, S., Schmid, C.: Local features and kernels for classification of texture and object categories: A comprehensive study. Int. J. Comput. Vis. **73**(2), 213–238 (2007)

Chapter 5
Description of Interest Regions

Local photometric descriptors computed for regions around interest points have been very successful in many problems. These local features are distinctive, and robust with respect to changes in viewpoint, scale and occlusion. The most widely used is the SIFT descriptor that uses gradient as the local feature [8]. This chapter describes how the center-symmetric LBP (CS-LBP) operator introduced in Chap. 2 can be used as the local feature in the SIFT algorithm, combining the strengths of the SIFT and LBP [4]. The resulting descriptor is called the CS-LBP descriptor.

5.1 Related Work

Local image feature detection and description have received much attention in recent years. The basic idea is to first detect interest regions that are covariant to a class of transformations. Then, an invariant descriptor is built for each detected region. After computing the descriptors, interest regions between images can be matched. This approach has many advantages, because local features can be made very tolerant to illumination changes, perspective distortions, image blur, image zoom etc. The approach is also robust to occlusions. Local features have performed very well in many computer vision applications, such as image retrieval [9], wide baseline matching [16], object recognition [8], texture recognition [7], and robot localization [15].

The interest regions that are used as input to region description methods are provided by the interest region detectors. Many different approaches to region detection have been proposed. For example, some detectors detect corner-like regions while others extract blobs. Since this chapter focuses on interest region description, the reader is referred to [13] for more information on interest region detection.

As with the interest region detection, many different approaches to interest region description have been proposed. The methods emphasize different image properties such as pixel intensities, color, texture, and edges. Many of the proposed descriptors are distribution-based, i.e. they use histograms to represent different characteristics of appearance or shape. Among the best known methods are the *intensity-domain*

M. Pietikäinen et al., *Computer Vision Using Local Binary Patterns*,
Computational Imaging and Vision 40,
DOI 10.1007/978-0-85729-748-8_5, © Springer-Verlag London Limited 2011

spin image [7], the *SIFT* descriptor [8], the *GLOH* descriptor [12], the *shape context* [12], and the *SURF* descriptor [1]. There exist many comparative studies on region descriptors [3, 12, 14]. Almost in every study, the best results are reported for distribution-based descriptors such as SIFT.

Many existing texture operators have not been used for describing interest regions. One reason might be that, by using these methods, usually a large number of dimensions is required to build a reliable descriptor. The LBP has properties that favor its usage in interest region description such as tolerance against illumination changes and computational simplicity. Drawbacks are that the operator produces a rather long histogram and is not too robust on flat image areas. To address these problems, a new LBP-based texture feature was proposed [4], denoted as *center-symmetric local binary pattern* (CS-LBP) that is more suitable for the given problem. The CS-LBP operator is described in Chap. 2.

Because the SIFT and other related distribution-based descriptors [1, 2, 12] have shown excellent performance in different problems, the focus of the presented work was on this approach. It was of special interest to see if the gradient orientation and magnitude based feature used in the SIFT algorithm could be replaced by a different feature that obtains better or comparable performance. For this purpose, a new interest region descriptor was developed, denoted as *CS-LBP descriptor* that combines the good properties of the SIFT and LBP. It is achieved by adopting the SIFT descriptor and using the CS-LBP feature instead of original gradient feature. This feature allows simplifications of several steps of the algorithm which makes the resulting descriptor computationally simpler than SIFT. It also appears to be more robust against illumination changes than the SIFT descriptor. Before presenting in detail the CS-LBP descriptor, a brief review of the SIFT method that forms another basis for the CS-LBP is given.

The SIFT descriptor is a 3D histogram of gradient locations and orientations. Location is quantized into a 4×4 location grid and the gradient angle is quantized into eight orientations, resulting in a 128-dimensional descriptor. First, the gradient magnitudes and orientations are computed within the interest region. The gradient magnitudes are then weighted with a Gaussian window overlaid over the region. To avoid boundary effects in the presence of small shifts of the interest region, a trilinear interpolation is used to distribute the value of each gradient sample into adjacent histogram bins. The final descriptor is obtained by concatenating the orientation histograms over all bins. To reduce the effects of illumination change the descriptor is first normalized to unit length. Then, the influence of large gradient magnitudes is reduced by thresholding the descriptor entries, such that each one is no larger than 0.2, and renormalizing to unit length.

5.2 CS-LBP Descriptor

In the following, the CS-LBP descriptor is presented in detail. The input for the descriptor is a normalized interest region. The process is depicted in Fig. 5.1 [4]. In

(a) Detected Hessian–Affine Region   **(b) Normalized Region with Location Grid**

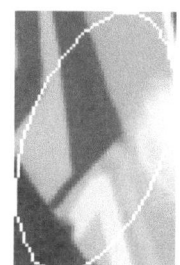   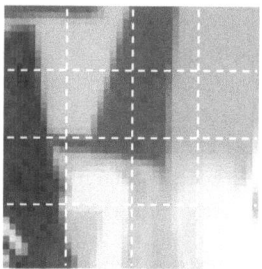

(c) CS–LBP Descriptor for the Normalized Region

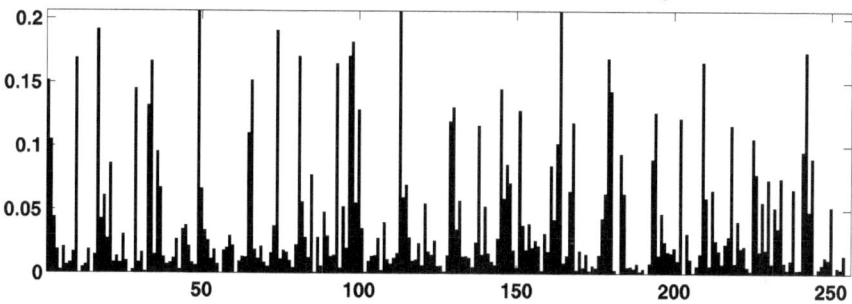

Fig. 5.1 The CS-LBP descriptor. (**a**) An elliptical image region detected by Hessian-Affine detector. (**b**) The region with Cartesian location grid after affine normalization. (**c**) The resulting CS-LBP descriptor computed for the normalized region

the experiments, the region size after normalization is fixed to 41×41 pixels and the pixel values lie between 0 and 1.

First a feature for each pixel of the input region is extracted by using the CS-LBP operator introduced in Sect. 2.8. The operator has 3 parameters: radius R, number of neighboring pixels N, and threshold on the gray-level difference T. According to the experiments good values for these parameters are usually found from $\{1, 2\}$ for R, $\{6, 8\}$ for N, and $\{0, \ldots, 0.02\}$ for T.

A weight is associated with each pixel of the input region based on the used feature. A comparison of three different weighting strategies, namely *uniform*, *Gaussian-weighted gradient magnitude (SIFT)*, and *Gaussian*, showed that simple uniform weighting is the most suitable choice for the CS-LBP feature. In other words, the feature weighting step can be omitted in this case.

In order to incorporate spatial information into the descriptor, the input region is divided into cells with a location grid. In the experiments presented here, a 4×4 (16 cells) Cartesian grid is used. For each cell a CS-LBP histogram is built. Thus, the resulting descriptor is a 3D histogram of CS-LBP feature locations and values. As explained earlier, the number of different feature values ($2^{N/2}$) depends on the neighborhood size (N) of the chosen CS-LBP operator. In order to avoid boundary effects in which the descriptor abruptly changes as a feature shifts from one cell to another, bilinear interpolation over x and y dimensions is used to share the

weight of the feature between four nearest cells. The share for a cell is determined
by the bilinear interpolation weights. Interpolation over feature value dimension is
not needed because the CS-LBP feature is quantized by its nature.

The final descriptor is built by concatenating the feature histograms computed
for the cells to form a $M \times M \times 2^{N/2}$-dimensional vector, where the M and N are
the grid size and CS-LBP neighborhood size, respectively. For $(M = 4, N = 6)$, and
$(M = 4, N = 8)$ the lengths of the CS-LBP descriptors are 128, and 256, respectively. The descriptor is then normalized to unit length. The influence of very large
descriptor elements is reduced by thresholding each element to be no larger than
a threshold. This means that the distribution of features has greater emphasis than
individual large values. After empirical testing the threshold was fixed to 0.2 which
is exactly the same value as used in the SIFT algorithm. Finally, the descriptor is
renormalized to unit length.

5.3 Image Matching Experiments

Two well-known protocols were used to evaluate the CS-LBP descriptor. Both are
freely available on the Internet. The first protocol is a matching protocol that is
designed for matching interest regions between a pair of images [5]. The second
protocol is the *PASCAL Visual Object Classes Challenge 2006* protocol which is an
object category classification protocol [6]. The performance of the CS-LBP descriptor was compared to the state-of-the-art descriptor SIFT. This allows to evaluate how
the CS-LBP feature compares to the gradient one. Next, the interest region detection
methods used in the experiments are explained, and then the matching test protocol
and some results are described. For additional results and the object categorization
protocol, see [4].

The interest region detectors extract the regions which are used to compute the
descriptors. In the experiments of [4], four different detectors were used: *Hessian-
Affine (HesAff)*, *Harris-Affine (HarAff)*, *Hessian-Laplace (HesLap)*, and *Harris-
Laplace (HarLap)* [11, 13]. The HesAff and HarAff detectors output different types
of image structures. HesAff detects blob-like structures while HarAff looks for
corner-like structures. Both detectors output elliptic regions of varying size determined by the detection scale. Before computing the descriptors, the detected regions
are mapped to a circular region of constant radius to obtain scale and affine invariance. Rotation invariance, if required by the application, is obtained by rotating the
normalized regions in the direction of the dominant gradient orientation, as suggested in [8]. HesLap and HarLap are the scale invariant versions of the HesAff and
HarAff detectors, respectively. They differ from the affine invariant detectors in that
they omit the *affine adaptation* step [11]. In the experiments, the normalized region
size is fixed to 41×41 pixels. For region detection, normalization, and computing
SIFT descriptors the software routines provided by the matching protocol [5] were
used.

The image matching protocol used is available on the Internet together with the
test data [5]. The test data includes images with different geometric and photometric

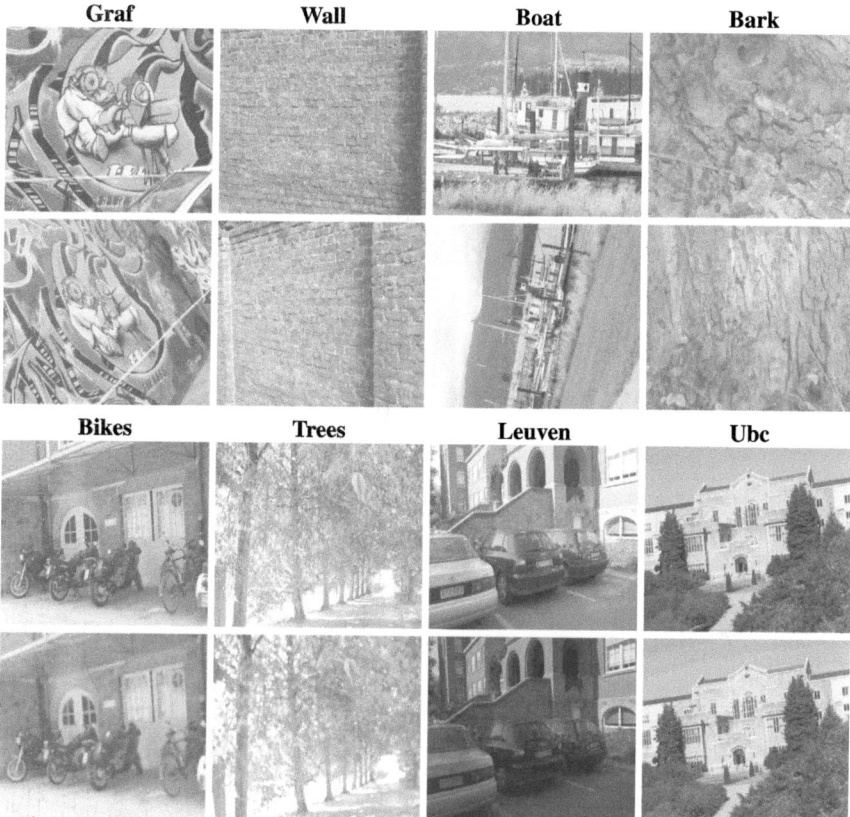

Fig. 5.2 Test images: **Graf** (viewpoint change, structured scene), **Wall** (viewpoint change, textured scene), **Boat** (scale change + image rotation, structured scene), **Bark** (scale change + image rotation, textured scene), **Bikes** (image blur, structured scene), **Trees** (image blur, textured scene), **Leuven** (illumination change, structured scene), and **Ubc** (JPEG compression, structured scene)

transformations and for different scene types. Six different transformations were evaluated: *viewpoint change*, *scale change*, *image rotation*, *image blur*, *illumination change*, and *JPEG compression*. The two different scene types are *structured* and *textured* scenes. These test images are shown in Fig. 5.2 [4, 5]. The images are either of planar scenes or the camera position was fixed during acquisition. The images are, therefore, always related by a homography (included in the test data). In order to study in more detail the tolerance of the descriptor to illumination changes, four additional image pairs shown in Fig. 5.3 were captured [4].

The evaluation criterion is based on the number of correct and false matches between a pair of images. The definition of a match depends on the matching strategy. As in [12], two interest regions were declared to be matched if the Euclidean distance between their descriptors is below a threshold. The number of correct matches is determined with the *overlap error* [10]. It measures how well the regions A and B correspond under a known homography H, and is defined by the ratio of the inter-

IC1   IC2   IC3   IC4

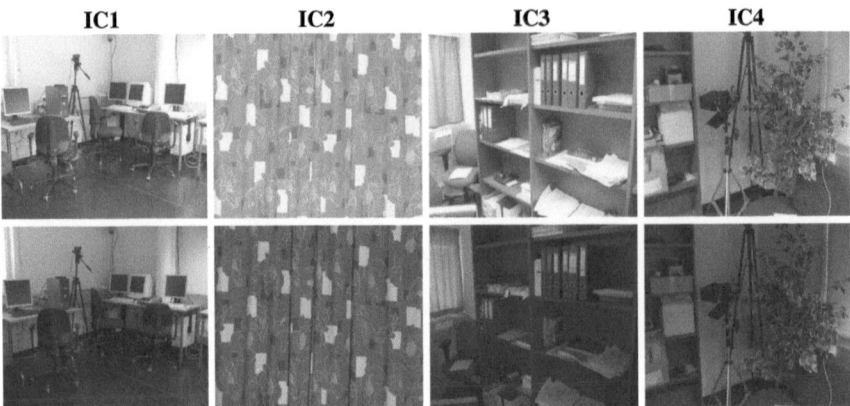

Fig. 5.3 Additional test images for illumination changes

section and union of the regions: $\varepsilon_S = 1 - (A \cap H^T B H)/(A \cup H^T B H)$. A match is assumed to be correct if $\varepsilon_S < 0.5$. A descriptor can have several matches and several of them may be correct. The results are presented with *recall* versus *1-precision*:

$$recall = \frac{\#correct\ matches}{\#correspondences}, \qquad 1\text{-}precision = \frac{\#false\ matches}{\#all\ matches}, \qquad (5.1)$$

where the #*correspondences* stands for the ground truth number of matching regions between the images. The curves are obtained by varying the distance threshold and a perfect descriptor would give a recall equal to 1 for any precision.

5.3.1 Matching Results

As an example of the extensive experiments presented in [4], Fig. 5.4 shows matching results for HesAff and HarAff regions. The ranking between the two descriptors seems to be more or less invariant to region detection method. Better overall performance is obtained with the HesAff regions. The results show that for most of the test cases the CS-LBP descriptor is able to outperform the SIFT descriptor. For the rest of the cases, comparable performance is achieved. In the case of illumination changes, i.e. for the *Leuven* and *IC1-IC4* sequences, CS-LBP seems to be more robust. This could be explained by the fact that by its nature LBP is invariant with respect to monotonic gray scale transformations at pixel level while in SIFT normalization is done at region level.

In the previous experiments, the CS-LBP$_{2,8,0.01}$ operator used has a length 256, i.e. twice as long as the SIFT descriptor. In order to make the descriptor dimension the same as for the SIFT descriptor, the neighborhood size of the CS-LBP operator was decreased from 8 to 6. It was shown in [4] that there were only slight changes in the results when compared to the ones with longer CS-LBP descriptor.

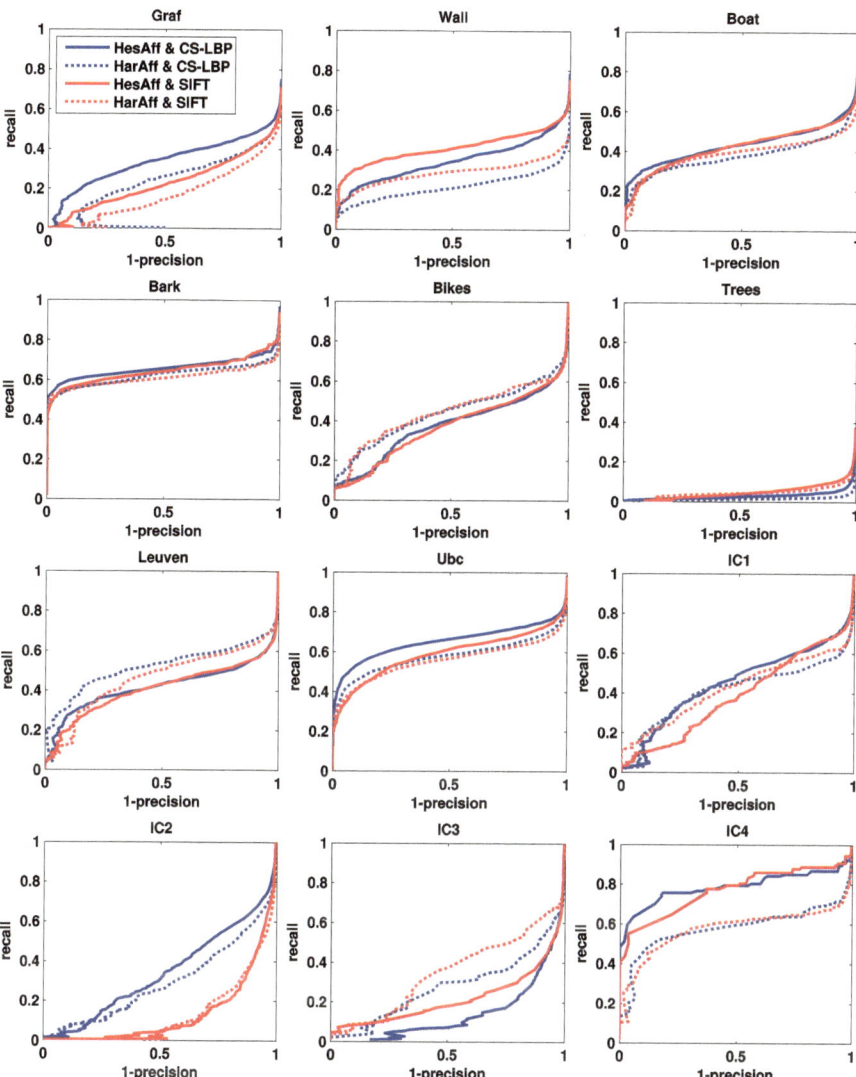

Fig. 5.4 Matching results for HesAff and HarAff regions. The CS-LBP descriptor uses 4×4 grid and the CS-LBP$_{2,8,0.01}$ with uniform weighting

5.4 Discussion

The CS-LBP descriptor combines the strengths of two powerful methods, the SIFT descriptor and the LBP texture operator. It uses a SIFT-like grid and replaces SIFT's gradient features with an LBP-based CS-LBP feature. The CS-LBP feature has properties that make it well suited for this task, including a relatively short feature histogram, robustness to illumination changes, and computational simplicity. Furthermore, it does not require many parameters to be set. The performance of the

CS-LBP descriptor has been compared to that of the SIFT descriptor in the contexts of matching and object category classification [4]. The CS-LBP has provided better performance in the matching experiments while equal performance was obtained for object category classification. One possible explanation might be that the discriminative power of the descriptor, which is very important in image matching where exact correspondences are needed, is less important in the context of category classification where similar regions without a need for exact matches are looked for.

References

1. Bay, H., Tuytelaars, T., Gool, L.J.V.: SURF: Speeded up robust features. In: Proc. European Conference on Computer Vision, vol. 1, pp. 404–417 (2006)
2. Belongie, S.J., Malik, J., Puzicha, J.: Shape matching and object recognition using shape contexts. IEEE Trans. Pattern Anal. Mach. Intell. **24**(4), 509–522 (2002)
3. Carneiro, G., Jepson, A.D.: Multi-scale phase-based local features. In: Proc. IEEE Conference on Computer Vision and Pattern Recognition, vol. 1, p. 736 (2003)
4. Heikkilä, M., Pietikäinen, M., Schmid, C.: Description of interest regions with local binary patterns. Pattern Recognit. **42**(3), 425–436 (2009)
5. http://www.robots.ox.ac.uk/~vgg/research/affine/
6. http://www.pascal-network.org/challenges/VOC/voc2006
7. Lazebnik, S., Schmid, C., Ponce, J.: A sparse texture representation using local affine regions. IEEE Trans. Pattern Anal. Mach. Intell. **27**(8), 1265–1278 (2005)
8. Lowe, D.G.: Distinctive image features from scale-invariant keypoints. Int. J. Comput. Vis. **60**, 91–110 (2004)
9. Mikolajczyk, K., Schmid, C.: Indexing based on scale invariant interest points. In: Proc. International Conference on Computer Vision, vol. 1, pp. 525–531 (2001)
10. Mikolajczyk, K., Schmid, C.: An affine invariant interest point detector. In: Proc. European Conference on Computer Vision, pp. 128–142 (2002)
11. Mikolajczyk, K., Schmid, C.: Scale and affine invariant interest point detectors. Int. J. Comput. Vis. **60**(1), 63–86 (2004)
12. Mikolajczyk, K., Schmid, C.: A performance evaluation of local descriptors. IEEE Trans. Pattern Anal. Mach. Intell. **27**(10), 1615–1630 (2005)
13. Mikolajczyk, K., Tuytelaars, T., Schmid, C., Zisserman, A., Matas, J., Schaffalitzky, F., Kadir, T., Gool, L.V.: A comparison of affine region detectors. Int. J. Comput. Vis. **65**(1/2), 43–72 (2005)
14. Pierre, M., Pietro, P.: Evaluation of features detectors and descriptors based on 3D objects. Int. J. Comput. Vis. **73**(3), 263–284 (2007)
15. Se, S., Lowe, D., Little, J.: Global localization using distinctive visual features. In: IEEE/RSJ International Conference on Intelligent Robots and Systems, vol. 1, pp. 226–231 (2002)
16. Tuytelaars, T., Van Gool, L.J.: Matching widely separated views based on affine invariant regions. Int. J. Comput. Vis. **59**(1), 61–85 (2004)

Chapter 6
Applications in Image Retrieval and 3D Recognition

This chapter introduces two applications using spatial domain LBP. First, two variants of a block-based method for content-based image retrieval (CBIR) are described [24]. These methods provided very promising results in retrieving images from the Corel database, for example. Then, a method for recognizing 3D textured surfaces using multiple LBP histograms is presented [18]. This approach performed very well in the classification of Columbia-Utrecht database (CUReT) textures imaged under different viewpoints and illumination directions. It also provided very promising results in the classification of outdoor scene images.

6.1 Block-Based Methods for Image Retrieval

Content-based image retrieval (CBIR) has received wide interest in recent years. Due to the growing number of image and video databases in the Internet and other information sources, better and better retrieval methods are needed.

The most common image descriptors used for retrieval are based on color, texture and shape information. The most widely used color features include color histograms [22], color correlograms [8], color moments [21] and MPEG-7 color descriptors [12]. Many different texture descriptors have also been used. Some approaches like two of the MPEG-7 texture descriptors [12] are based on Gabor filtering [11]. Other texture features used include DFT transformation [20], features based on visual perception [25], and the block-based edge histogram descriptor [17] included in the MPEG-7 standard.

LBP and its derivatives have been successfully used in some CBIR systems, but the use of the operator has been often limited to the original version [13] and it has been applied on full images only [29].

The CBIR texture descriptors used in commercial systems have been often calculated for full images. This approach is well justified because it usually keeps the size of the feature database reasonably low, but of course this depends on the used features and the number of images. But there is still a problem when considering only full images. The local image areas of interest are often left unnoticed as the

M. Pietikäinen et al., *Computer Vision Using Local Binary Patterns*,
Computational Imaging and Vision 40,
DOI 10.1007/978-0-85729-748-8_6, © Springer-Verlag London Limited 2011

global features do not contain enough information for local discrimination. A possible way to deal with local properties is to use image segmentation. However, the segmentation is usually prone to errors making it not very suitable for images with general unknown content. Another way to improve retrieval performance is to apply the feature extractors to subimage areas without using any segmentation and to compare the obtained feature descriptors separately.

In this section, two block-based texture methods for content-based image retrieval (CBIR) are described [24]. In the first approach the query and database images are divided into equally sized blocks from which LBP histograms are extracted. Then the block histograms are compared using a relative L1 dissimilarity measure based on the Minkowski distances. The second approach uses the image division on database images and calculates a single feature histogram for the query. It sums up the database histograms according to the size of the query image and finds the best match by exploiting a sliding search window. The first method was evaluated against color correlogram and edge histogram based algorithms. The second, user interaction dependent approach is used to provide example queries. The experiments showed very promising results.

6.1.1 Description of the Method

6.1.1.1 Nonparametric Dissimilarity Measure

A distance function is needed for comparing images on the basis of their LBP feature distributions. Many different dissimilarity measures [19] have been proposed. Most of the LBP studies have favored a nonparametric log-likelihood statistic as suggested by Ojala et al. [14]. In this study, however, a relative L1 measure similar to the one proposed by Huang et al. [8] was chosen due to its performance in terms of both speed and good retrieval rates when compared to the log-likelihood and other available statistics. In the initial tests the log-likelihood and relative L1, which were clearly better than the rest, produced even results but the calculation of relative L1 measure took only a third of the time required by log-likelihood. The dissimilarity measure is given in Eq. 6.1, where $x1$ and $x2$ represent the feature histograms to be compared and subscript i is the corresponding bin.

$$L_1^{relative}(x_1, x_2) = \sum_i \frac{|x_{1,i} - x_{2,i}|}{x_{1,i} + x_{2,i}}. \tag{6.1}$$

6.1.1.2 The Block Division Method

The block division method is a simple approach that relies on subimages to address the spatial properties of images. It can be used together with any histogram descriptors similar to LBP. The method works in the following way: First it divides the model images into square blocks that are arbitrary in size and overlap. Then the method calculates the LBP distributions for each of the blocks and combines the histograms into a single vector of sub-histograms representing the image. In the query phase the same is done for the query image(s) after which the query and model are compared by calculating the distance between each sub-histogram of the query and model. The final image dissimilarity D for classification is the sum of minimum distances as presented by Eq. 6.2:

$$D = \sum_{i=0}^{N-1} \min_{j}(D_{i,j}), \tag{6.2}$$

where N is the total amount of query image blocks and $D_{i,j}$ the distance (relative L1) between the ith query and jth model block histograms. An example of the approach in operation is shown in Fig. 6.1 [24]. Note that in this figure the shown histograms are only examples and not the actual LBP distributions of corresponding image blocks.

6.1.1.3 The Primitive Blocks Method

Another way to utilize image blocks is to use small constant-sized elements referred here as primitive blocks. Instead of larger and less adaptive equivalents, the primitive blocks can be combined to match the size of the query image with reasonable accuracy and speed as there is no heavy processing involved like in the pixel-by-pixel sliding window methods. In this approach the model images are handled as in the previous method but the query images are left untouched and only a global LBP histogram is produced for each of them. The model's sub-histograms H_i are summed up to a single feature histogram according to

$$H = \sum_{i=0}^{N-1} H_i, \tag{6.3}$$

by first adapting to the size of the query image, and then they are normalized. The primitive blocks (actually the corresponding block histograms) are connected in the way depicted in Fig. 6.2 [24], where the search window goes through the whole model image and does the matching by using the distance measure of Eq. 6.1. The size of the search window is the same or a bit larger, depending on the chosen block size, than the area dictated by the query image dimensions.

While using primitive blocks, there exist two types of overlapping. The blocks themselves can overlap, as in the case of previous block division method, and then the measure of overlapping is determined in single pixels. The same applies to the search window areas consisting of primitive blocks but in their case the overlap is quantified to the dimensions of the used primitive blocks.

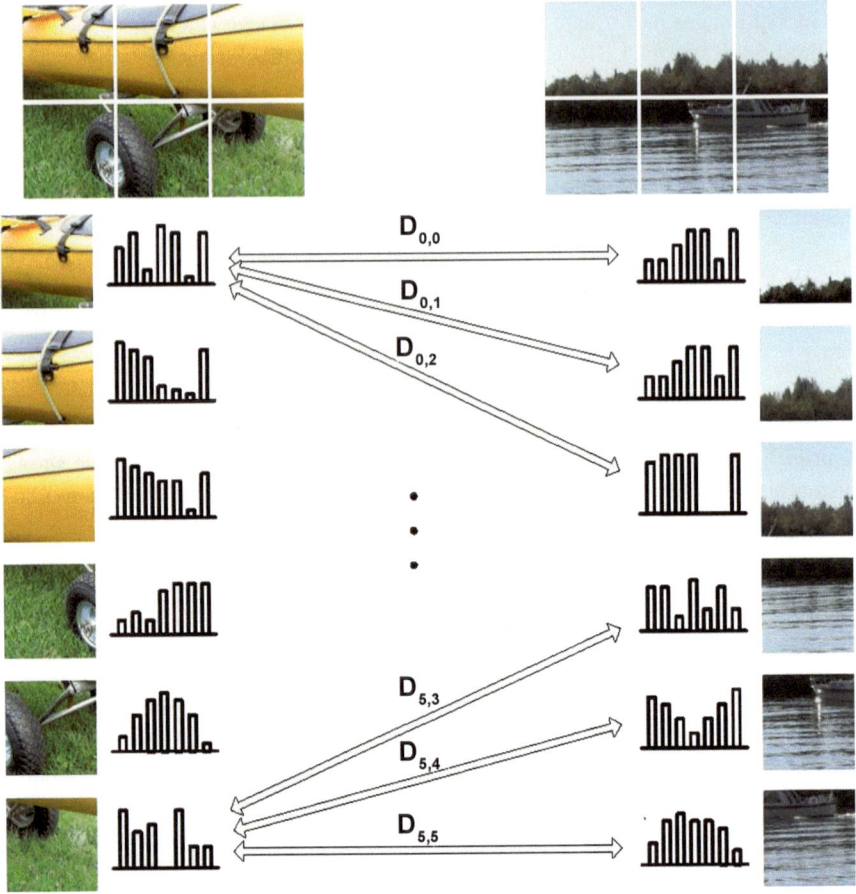

Fig. 6.1 The block division method

6.1.2 Experiments

6.1.2.1 Test Database and Configurations

Both image retrieval methods were tested on a database consisting of commercial
Corel Image Gallery [3] images of sizes 384×256 and 256×384. The image categorization was set according to the original image database structure of the Corel
set, so there were 27 categories of 50 images each making up 1350 images in total.
No further categorization was utilized. This kind of categorization may sound rude
but it was used to ensure the reproducibility of the tests. The following categories
(physical database folder names) where chosen from the image gallery: Apes, Bears,
Butterflies, Cards, Death Valley, Dogs, Elephants, Evening Skies, Fancy Flowers,
Fireworks, Histology, Lighthouses, Marble Textures, Night Scenes, Owls, Rhinos
and Hippos, Roads and Highways, Rome, Skies, Snakes Lizards and Salamanders,

$$H = H_0 + H_1 + H_2 + \dots + H_{N-1}$$

Fig. 6.2 The primitive blocks approach

Fig. 6.3 Image examples from the Corel Gallery database

Space Voyage, Sunsets Around the World, Tigers, Tools, Waterscapes, Wildcats, and Winter. Some example images are shown in Fig. 6.3 [24].

The category experiments were carried on five different image categories (Apes, Death Valley, Fireworks, Lighthouses, and Tigers), so there were 250 queries per experiment. Two different image feature descriptors, one based on color and the other one on texture, were chosen to be compared to the queries attained with LBP operators. The first one of them was the color correlogram [8], which is still one of the most powerful color descriptors, and the other one was the edge histogram [17] that operates with image blocks and is included in the MPEG-7 Visual Standard [12]. The correlogram was applied with four distances (1, 3, 5, and 7) and four quantization levels per color channel, that is 64 quantization levels in total. The edge histogram used the standard parameters as used by Park et al. [16], thus the method produced histograms of 80 bins.

LBP was applied both on full images and image blocks of sizes 128×128 and 96×96. Two clearly different LBP operators were tried out: one using eight uninterpolated samples and a predicate of 1 ($\text{LBP}_{8,1}^{u2}$) and a multiresolution version

Table 6.1 The results (precision/recall) for different methods

Method	Block size/overlap	10 images (%)	25 images	50 images
Color correlogram	full image	37.8/7.5	25.3/12.7	18.7/18.7
Edge histogram	image dependent	26.3/5.3	18.4/9.2	14.2/14.2
$LBP^{u2}_{8,1}$	full image	34.7/6.9	24.6/12.3	19.0/19.0
$LBP^{u2}_{8,1+8,2.4+8,5.4}$	full image	36.9/7.5	25.3/12.7	18.7/18.7
$LBP^{u2}_{8,1}$	$128 \times 128\ (0 \times 0)$	36.9/7.4	26.5/13.2	21.3/21.3
$LBP^{u2}_{8,1+8,2.4+8,5.4}$	$128 \times 128\ (0 \times 0)$	37.4/7.5	26.6/13.3	20.4/20.4
$LBP^{u2}_{8,1}$	$128 \times 128\ (64 \times 64)$	43.0/8.6	31.3/15.7	23.7/23.7
$LBP^{u2}_{8,1+8,2.4+8,5.4}$	$128 \times 128\ (64 \times 64)$	43.3/8.7	31.0/15.5	24.0/24.0
$LBP^{u2}_{8,1}$	$96 \times 96\ (0 \times 0)$	40.5/8.1	29.0/14.5	22.5/22.5
$LBP^{u2}_{8,1+8,2.4+8,5.4}$	$96 \times 96\ (0 \times 0)$	41.0/8.2	29.2/14.6	21.5/21.5
$LBP^{u2}_{8,1}$	$96 \times 96\ (48 \times 48)$	45.5/9.1	31.6/15.8	24.0/24.0
$LBP^{u2}_{8,1+8,2.4+8,5.4}$	$96 \times 96\ (48 \times 48)$	45.3/9.0	32.5/16.2	23.4/23.4

with three different radii and eight interpolated samples in each ($LBP^{u2}_{8,1+8,2.4+8,5.4}$). Both operators relied on uniform patterns with U value of 2 ($u2$), so the corresponding histograms had 59 and 177 bins, respectively.

6.1.2.2 Results

The results of the experiments are shown in Table 6.1. The used measures are precision (the ratio between the correct and all retrieved images) and recall (the ratio between the correct retrieved images and all correct images in the database), and the numbers are marked as percentages. The LBPs using overlapping blocks achieved precision rates of about 45% (over 9% for recall) for 10 images and are clearly better than any of the other extractors. Their results for 50 images are almost as good as those obtained with the best full image operators for 25 images. Using blocks, especially overlapping ones, instead of full images seems to make a clear difference. It is also to be noticed that using overlap has a large impact regardless of the block size. Several percentages are gained through 50% overlapping.

The test with the primitive block approach was performed with an $LBP^{u2}_{8,1}$ operator without interpolation. Figure 6.4 shows an example query obtained by using 32×32 sized primitive blocks with overlap of two pixels (the overlap between search windows was set to 50%) [24]. The query images have been taken from an original database image and they have been outlined in such a way that they are composed of mostly homogeneous texture. The actual retrieval task was not an easy one: two subimages were needed to get satisfying results, that is seven out of 16 images from the right category (Lighthouses). The chosen subimages have both very distinctive appearance but the image of a rock appeared to have more positive influence on the outcome than the carefully outlined picture of the white lighthouse

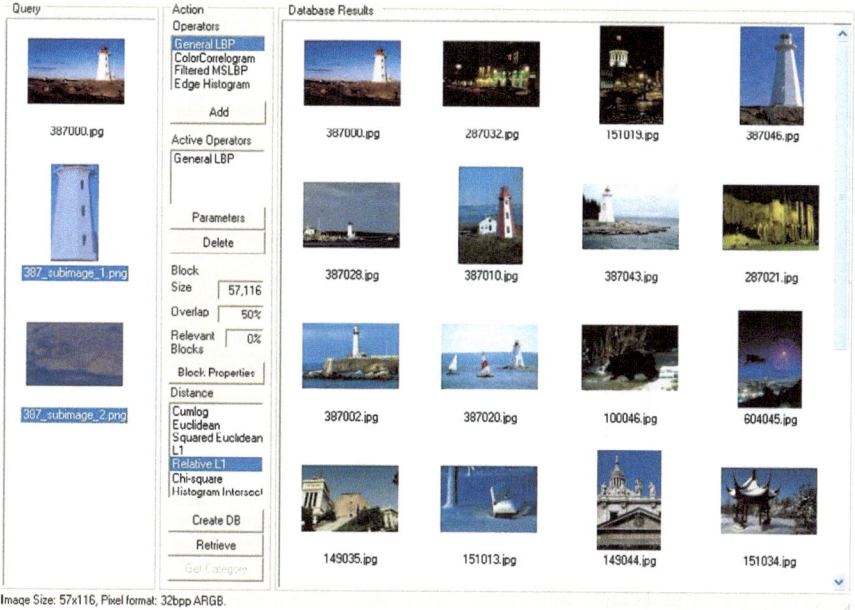

Image Size: 57x116, Pixel format: 32bpp ARGB.

Fig. 6.4   A test query. The *left box* of images in the user interface is the query group from which only the outlined (darkened) subimages were selected for the eventual search

itself. The probable reason for this is the clear distinction in the properties of the two subimages—the image of the rock is rich in its texture content.

Some additional tests were also conducted with a stamp database consisting of about 900 German stamp images [23]. A couple of LBP extractors were used and their performance was evaluated against a commercial stamp image retrieval software. The block division method fared at least as well or even better than the matured retrieval application making use of multiple different image features (color, texture, motive, and image size and aspect ratios).

6.1.3 Discussion

In this section, the use of LBP texture features combined with two different block-based image division methods was considered. The results obtained show that the LBP can be successfully used to retrieve images with general content as it is fast to extract and it has useful qualities like invariance to monotonic transitions in gray scale and small descriptor size. The color correlogram, that represents the state of the art in CBIR, was clearly outperformed by one of the developed subimage approaches.

The increased retrieval rates of the tested methods come at the expense of higher computational demands. The time needed for query grows linearly with the amount of used image blocks. With large images and small block sizes the required processing capacity slips easily out of the grasp of applications that have real-time requirements. Still, it should be noted that it does not seem to be necessary to use large numbers of small blocks as, according to the obtained results, a few blocks per image is usually enough to make a considerable difference when compared to descriptors calculated for full images.

The method based on primitive blocks was hard to assess as there is a level of user interaction involved in the query procedure. Nevertheless, it has some important properties that increase its value in the field of CBIR: It is faster than conventional search window approaches as it does not extract features for every possible search window size separately. Another noteworthy feature is that it can be used to find objects consisting of a single texture or larger entities with several different areas of interest as the query can be adjusted by using more than one sample image.

For the future studies, there are many ways that could enhance the performance and speed of the studied methods. For instance, different block matching algorithms, like the three-step search method, could be used to speed up the matching process. Another possibility could be to use image blocks that are weighted according to their spatial positions. In the case of multiresolution LBP, the use of weights could be extended to emphasize the LBPs containing the most relevant texture information. These and other enhancements could improve the usability of LBP features in the CBIR of the future.

Another example on the use of LBP in content-based retrieval is by Grangier and Bengio [7], who developed a discriminative model for the retrieval of images from text queries. They extracted visual features by dividing each picture into overlapping square blocks, and each block was then described with edge and color histograms. Uniform LBPs were used for computing the edge histograms.

6.2 Recognition of 3D Textured Surfaces

The analysis of 3D textured surfaces has been a topic of considerable interest in recent years due to many potential applications, including classification of materials and objects from varying viewpoints, classification and segmentation of scene images for outdoor navigation, aerial image analysis, and retrieval of scene images from databases.

Due to the changes in viewpoint and illumination, the visual appearance of different surfaces can vary greatly, which makes their recognition very difficult. The simplest solution to the recognition of 3D textured objects is to apply 2D texture analysis methods "as they are" [2].

Malik et al. proposed a method based on learning representative texture elements (textons) by clustering, and then describing the texture by their distribution [10]. The image is first processed with a multichannel filter bank. Then, the filter responses are clustered into a small set of prototype response vectors, i.e. textons. The vocabulary of textons corresponds to the dominant features in the image: bars and edges at various orientations and phases. Each pixel in the texture gets the label of the best matching texton. The histogram of the labels computed over a region is used for texture description. Leung and Malik extended this approach to 3D surfaces by constructing the vocabulary of 3D textons [9] and demonstrating the applicability of the method in the classification of Columbia-Utrecht database (CUReT) textures [6] taken in different views and illuminations.

Recent findings from human psychophysics, neurophysiology and computer vision provide converging evidence for a framework in which objects and scenes are represented as collections of viewpoint-specific local features rather than two-dimensional templates, or three-dimensional models [1].

Cula and Dana [4, 5] and Varma and Zisserman [27] used histograms of 2D textons extracted from training samples in different viewing and illumination conditions as texture models instead of determining 3D textons. As an alternative to the texton-based approach, the models used for texture description can also be built by quantizing the filter responses into bins and then normalizing the resultant histogram [28]. Based on the results of these studies, a robust view-based classification of 3D textured surfaces from a single image acquired under unknown viewpoint and illumination seems to be feasible. Using rotation-invariant features computed at three different scales (MR8 filter bank) Varma and Zisserman were able to classify all 61 CUReT textures with an accuracy of 96%, when 46 models for each texture class were used [27]. A problem with the proposed approaches is that the methods need many parameters to be set and are computationally complex, requiring learning of a representative texton library using e.g. K-means clustering, intensity normalization of gray scale samples e.g. to have zero mean and unit standard deviation, feature extraction by a multiscale filter bank, normalization of filter responses, and vector quantization of the multidimensional feature data to find the textons.

In the approach presented in this chapter 3D textures are modeled with multiple histograms of micro-textons, instead of more macroscopic textons used in earlier studies [18]. The micro-textons are extracted with the local binary pattern operator. This provides several advantages compared to the other methods. The performance of the approach was first assessed with the same CUReT textures that were used by in the earlier studies. In addition, classification experiments with a set of outdoor scene images were also carried out.

6.2.1 Texture Description by LBP Histograms

Varying lighting conditions and viewing angles greatly affect the gray scale properties of an image due to effects such as shading, shadowing or local occlusions. Therefore, it is important to use features, which are invariant with respect to gray scale changes. The textures may also be arbitrarily oriented, which suggests using rotation-invariant features. Due to foreshortening and other geometric transformations in a 3D environment, invariance to affine transformations should also be considered.

In the experiments presented in this section, neighborhoods with 8, 16 and 24 samples and radii 1, 3 and 5 were considered. The rotation-dependent operators chosen were LBP$_{8,1}$ (8 samples, radius 1), and multiresolution LBP$_{8,1+16,3+24,5}$ obtained by concatenating histograms produced by operators at three resolutions into a single histogram. In order to reduce the number of bins needed, the "uniform" pattern approach was used in the rotation-dependent case for the radii 3 and 5. The rotation-invariant operators LBP$_{8,1}^{riu2}$ and LBP$_{8,1+16,3+24,5}^{riu2}$ were also used, in which the matches of similar uniform patterns at different orientations are collected into a single bin.

LBP can be regarded as a "micro-texton" operator. At each pixel, it detects the best matching local binary pattern representing different types of (slowly or deeply sloped) curved edges, spots, flat areas etc. For LBP$_{8,1}^{riu2}$ operator, for example, the length of the feature vector and size of the texton vocabulary is as low as 10 (9 + 1 for "miscellaneous"). After scanning the whole image to be analyzed with the chosen operator, each pixel will have a label corresponding to one texton in the vocabulary. The histogram of the labels computed over a region is then used for texture description. For multiscale operators, the texton histograms computed at different scales are concatenated into a single histogram containing, for example, 54 textons ($= 10 + 18 + 26$) in case of the LBP$_{8,1+16,3+24,5}^{riu2}$ operator.

6.2.2 Use of Multiple Histograms as Texture Models

The 3D appearance of textures can be efficiently described with histograms of textons computed from training images taken from different views and illumination conditions. In the basic method presented here, one model histogram for each training sample is used. An unknown sample is then classified by comparing its texton histogram to the models of all training images, and the sample is assigned to the class of the nearest model.

LBP histograms were used as texture models. The histograms were normalized with respect to image size variations by setting the sum of their bins to one. For comparing histograms, a log-likelihood statistic was used:

$$L(S, M) = \sum_{n=0}^{N-1} S_b \log M_b, \qquad (6.4)$$

where b is the number of bin and S_b and M_b correspond to the sample and model probabilities at bin b, respectively.

Training of a system can be very problematic if a good coverage of training samples taken from different viewing angles and illumination conditions is required. Collecting such a large training set may even be impossible in real-world applications. A fundamental assumption in view-based vision is that each object of interest can be represented with a small number of models obtained from images taken from selected views. How many of these "keyframes" are needed and how they are selected is dependent on the data, features, and requirements of the given application.

For rough textures reasonably many models may be needed because the visual appearance of these textures can vary greatly, whereas smooth textures may require a much smaller number of models. By using invariant features (rotation, gray scale, affine) the within-class variability is likely to decrease, which should reduce the number of models needed. One should remember, however, that while adding feature invariance the discriminative power of a feature (and between-class differences) might in fact decrease.

How to select a good reduced set of models is dependent on the application. If a good coverage of training images of all classes taken from different viewpoints and illumination conditions is available and all classes are known beforehand, a model reduction method based on clustering or optimization can be used [27]. The method based on dimensionality reduction by principal component analysis considers only within-class variations [4, 5]. In the experiments originally presented in [18], the optimization approach was adopted in order to compare the results to the state-of-the art [27]. The optimization method is a simple hill-climbing search in which the set of samples is divided into two parts, one for training and the other for testing. Each sample in the training set is dropped off in turn, and a classification result against the test set is obtained. In each iteration, the sample whose removal results in the best classification accuracy is moved to the test set.

If it is not possible to have enough representative training samples, the selection of models could be done, for example, by utilizing information about the imaging positions and geometries of the objects, or temporal information available in image sequences. In [18], an approach for finding appearance models for each class by using self-organization was proposed.

6.2.3 Experiments with CUReT Textures

The CUReT database contains 61 different real-world textures, shown in Fig. 6.5 [18], each imaged under more than 200 different combinations of viewing and illumination directions [6]. In order to be able to compare the results, similar experiments as in earlier studies [4, 5, 27, 28] were carried out. In the first image set, 118 images taken under varying viewpoint and illumination with a viewing angle less than 60 degrees were extracted from each texture class in the database. The second set contained 156 images of each texture class with a viewing angle less than 70 degrees. Three different classification problems were considered containing 20, 40 and all 61 different textures. Half of the images (i.e. 59 or 78) for each texture were used for training and the other half for testing. The images for the training (and test) set were selected in two optional ways: by taking every alternate image, or by choosing them randomly and computing an average accuracy of 10 trial runs. 59 or 78 LBP histograms then modeled each texture, respectively, and in classification each sample from the test set was assigned to the class of the closest model histogram.

Table 6.2 shows the classification rates for different LBP features in each classification problem, when the viewing angle was less than 60 degrees and every alternate image was chosen for the training set. For comparison, the best corresponding results obtained by [27] with the MR8 filter bank are also presented. The results are

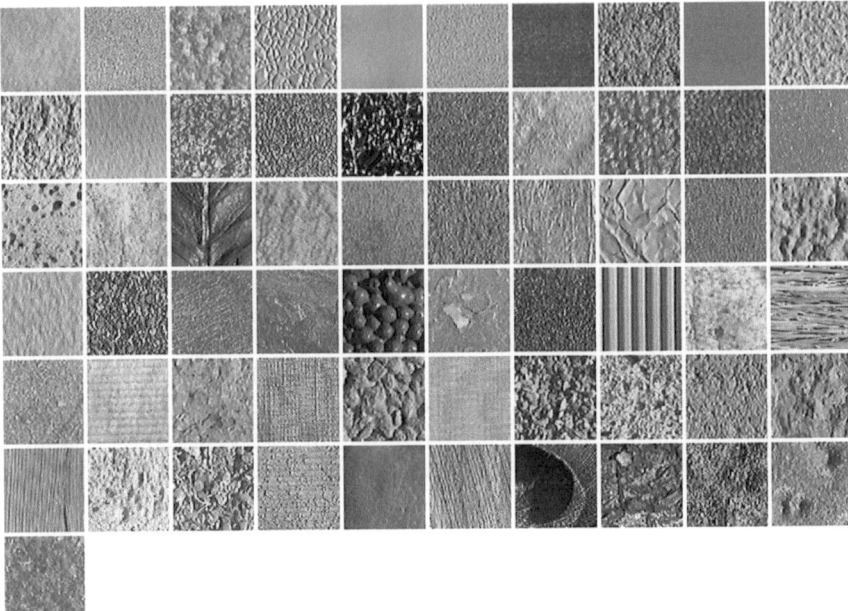

Fig. 6.5 CUReT textures

Table 6.2 Classification rates (%) for different number of texture classes

Operator	20 classes	40 classes	61 classes
$LBP_{8,1}$	97.54	91.57	87.02
$LBP_{8,1+16,3+24,5}$	98.73	94.49	90.03
$LBP_{8,1}^{riu2}$	93.73	83.69	81.47
$LBP_{8,1+16,3+24,5}^{riu2}$	98.81	97.25	96.55
MR8	97.50	96.30	96.07

not fully comparable, because the image sets were not exactly the same. Varma and Zisserman [27] used only 92 "sufficiently large" images for each class with a viewing angle less than 60 degrees, while all 118 images available with the same viewing angle limitation were used here. This classification problem is more challenging, because it included also small, often highly tilted samples, for which it is not easy to extract reliable feature statistics.

Multiresolution rotation-invariant LBP performed best, exceeding the classification rates obtained with the MR8 filter, even when a more difficult image set than in [27] was used.

The same operator achieved high classification rates also when the training samples were chosen randomly: 97.67%, 94.81% and 94.30% for the 20, 40 and 61 class problems, respectively. When using 156 images per class with viewing angle less than 70 degrees, and alternate sampling, very high rates of 97.63%, 95.2%

and 93.57% were still obtained for the given three problems. A large viewing angle means that highly tilted, often almost featureless samples were included in this image set.

It was also investigated how many models are needed for each class to obtain good performance when rotation-invariant LBP features are used. The problem of classifying 20 textures with a viewing angle less than 60 degrees was chosen for this study. The results presented in [18] show that excellent results can be obtained using only six models per texture, and also three models provided a satisfactory performance.

6.2.4 Experiments with Scene Images

Analysis of outdoor scene images, for example for navigating a mobile robot, is very challenging. Texture could play an important role in this kind of application, because it is much more robust than color with respect to changes in illumination and it could also be utilized in night vision [2]. Castano et al. argue that classification is a more important issue in a vast majority of applications, rather than clustering or unsupervised segmentation. In their work, they assessed the performance of two Gabor filtering based texture classification methods on a number of real-world images relevant to autonomous navigation on cross-country terrain and to autonomous geology. They obtained satisfactory results for rather simple terrain images containing four classes (soil, trees, bushes/grass, and sky), but the results for the rock dataset were much poorer.

For the experiments, a test set of outdoor images was created by taking a sequence of 22 color images of 2272×1704 pixels with a digital camera. A person was walking on a street and took a new picture after about every five meters. The camera was looking forward and its orientation with respect to the ground was kept roughly constant, simulating a navigating vehicle. The image set is available at the Outex database [15]. Half of the (gray-scale) images were used for training and the other half for testing. The images for training (and testing) were selected by taking every alternate image.

Five texture classes were defined: sky, trees, grass, road and buildings. Due to the considerable changes of illumination, the following sub-classes were also used: trees in the sun, grass in the sun, road in the sun, and buildings in the sun. Following the approach of Castano et al. [2], the ground-truth areas from each training and testing image were labeled by hand, i.e. areas in which a given texture class is dominating and not too close to another class to avoid border effects. Figure 6.6(a) shows an original (test) image and Fig. 6.6(b) a labeled image [18]. Next, each training image was processed with the chosen LBP operator and model distributions were created in two optional ways: (1) for each class and sub-class a single model histogram was computed using LBP-labeled pixels inside the ground-truth areas of the training set, (2) for each of the 11 images in the training set, a separate model histogram was created for those classes or sub-classes that were present in the given image. In the

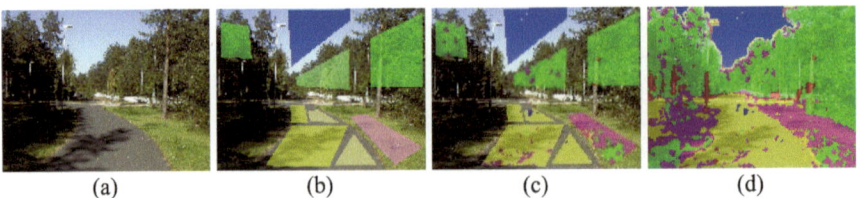

Fig. 6.6 A scene image. (a) The original image. (b) Ground-truth regions. (c) Classified pixels within ground-truth regions. (d) Segmented image

first case, the total number of model histograms was nine (in average 1.8 models per class), while in the second case it was 68 (in average 13.6 models per class).

Pixel-wise classification of the test images was performed by centering a circular disk with radius r $(r = 30)$ at the pixel being classified, computing the sample LBP histogram over the disk and assigning the pixel to the class whose model was most similar to the sample. After classifying all pixels inside the ground-truth regions in a similar way, classification rates for each of the five classes can be computed after combining classes with their possible sub-classes. Figure 6.6(c) shows an example of classified ground-truth pixels. Figure 6.6(d) demonstrates how the whole image could segmented by classifying all pixels in the image in a similar way.

Figure 6.7 presents the classification rates for the whole test set using different LBP operators with single or multiple histograms as models [18]. Rotation-variant multiresolution operator $LBP_{8,1+16,3+24,5}$ achieved a very good accuracy of 85.43%, but also the simple $LBP_{8,1}$ operator performed well (80.92%). Multiple histogram models were clearly better than single histograms. In this application area, the rotation-invariant LBP operators did not perform as well as their rotation-variant counterparts.

The scene textures used in the experiments had a wide variability within and between images due to variations in illumination, shadows, foreshortening, self-occlusion, and non-homogeneity of the texture classes. Therefore the results obtained can be considered very promising.

6.2.5 Discussion

The experiments presented in this section show that histograms of micro-textons provided by the multiscale LBP operator are very efficient descriptors of 3D surface textures imaged under different viewing and illumination conditions, providing excellent performance in the classification of CUReT textures. Due to the gray-scale and rotation invariance of the features used, only a few models per texture are needed for robust view-based recognition. The method performed well also in the classification of outdoor scene textures. These textures had a wide variability within and between images due to changes in illumination, shadows, foreshortening, self-occlusion, and non-homogeneity of the texture classes. Therefore the results obtained can be considered very promising, confirming that this approach has much

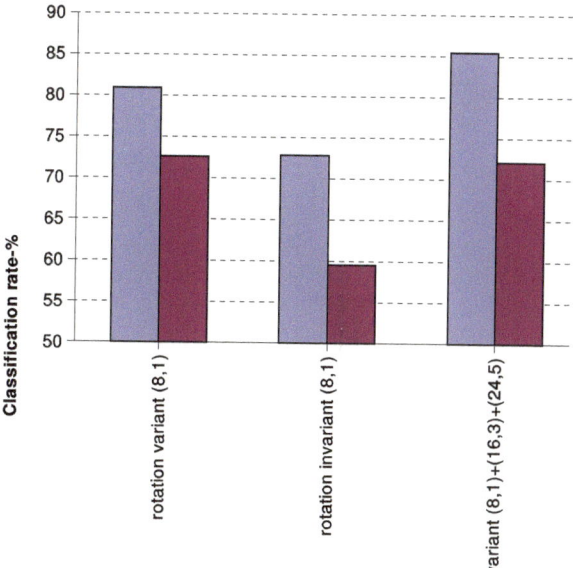

Fig. 6.7 Classification rates (%) for different versions of LBP. *Red* denotes single histograms and *blue* multiple histograms

potential for a wide variety of applications. After this study, an improved classification system combining local and contextual image texture information was proposed for scene classification [26]. Texture was modeled with LBP and classified locally with SVM. Another Conditional Random Field (CRF) based classifier was used to improve the results by taking into account contextual image constraints.

A significant advantage of the LBP-based method is that there is no need to create a specific texton library like in the earlier approaches, but a generic library of micro-textons can be used instead. Due to the invariance of the LBP features with respect to monotonic gray-scale changes, this method can tolerate considerable gray-scale variations common in natural images and no normalization of input images is needed. Unlike most other approaches, the proposed features are very fast to compute and do not require many parameters to be set. The only parameters needed in the LBP approach are the number of scales and the number of neighborhood samples.

The results suggest that the micro-textons detected by the LBP approach contain more discriminative texture information than the more coarsely-grained textons used in earlier studies. Micro-textons contain information about curved edges, spots and flat areas in a texture. These might be used as basic primitives for describing various types of textures, including microtextures, macrotextures, and non-homogeneous textures.

References

1. Bülthoff, H.H., Wallraven, C., Graf, A.: View-based dynamic object recognition based on human perception. In: Proc. International Conference on Pattern Recognition, pp. 768–776 (2002)
2. Castano, R., Manduchi, R., Fox, J.: Classification experiments on real-world texture. In: Proc. Workshop on Empirical Evaluation Methods in Computer Vision, pp. 3–20 (2001)
3. Corel Corporation (2005). http://www.corel.com/
4. Cula, O.G., Dana, K.J.: Compact representation of bidirectional texture functions. In: Proc. IEEE Conference on Computer Vision and Pattern Recognition, vol. 1, pp. 1041–1047 (2001)
5. Cula, O.G., Dana, K.J.: Recognition methods for 3D texture surfaces. In: Proc. SPIE Conference on Human Vision and Electronic Imaging, pp. 209–220 (2001)
6. Dana, K.J., van Ginneken, B., Nayar, S.K., Koenderink, J.J.: Reflectance and texture of real-world surfaces. ACM Trans. Graph. 18(1), 1–34 (1999)
7. Grangier, D., Bengio, S.: A discriminative kernel-based approach to rank images from text queries. IEEE Trans. Pattern Anal. Mach. Intell. 30(8), 1371–1384 (2008)
8. Huang, J., Kumar, S.R., Mitra, M., Zhu, W.J., Zabih, R.: Image indexing using color correlograms. In: Proc. IEEE Conference on Computer Vision and Pattern Recognition, pp. 762–768 (1997)
9. Leung, T., Malik, J.: Representing and recognizing the visual appearance of materials using three-dimensional textons. Int. J. Comput. Vis. 43(1), 29–44 (2001)
10. Malik, J., Belongie, S.J., Leung, T., Shi, J.B.: Contour and texture analysis for image segmentation. Int. J. Comput. Vis. 43(1), 7–27 (2001)
11. Manjunath, B.S., Ma, W.Y.: Texture features for browsing and retrieval of image data. IEEE Trans. Pattern Anal. Mach. Intell. 18, 837–842 (1996)
12. Manjunath, B.S., Ohm, J.R., Vinod, V.V., Yamada, A.: Color and texture descriptors. IEEE Trans. Circuits Syst. Video Technol. 11(6), 703–715 (2001). Special Issue on MPEG-7
13. Ojala, T., Pietikäinen, M., Harwood, D.: A comparative study of texture measures with classification based on feature distributions. Pattern Recognit. 29(1), 51–59 (1996)
14. Ojala, T., Pietikäinen, M., Mäenpää, T.: Multiresolution gray-scale and rotation invariant texture classification with local binary patterns. IEEE Trans. Pattern Anal. Mach. Intell. 24(7), 971–987 (2002)
15. Ojala, T., Mäenpää, T., Pietikäinen, M., Viertola, J., Kyllönen, J., Huovinen, S.: Outex—New framework for empirical evaluation of texture analysis algorithms. In: Proc. International Conference on Pattern Recognition, pp. 701–706 (2002)
16. Park, D.K., Jeon, Y.S., Won, C.S.: Efficient use of local edge histogram descriptor. In: Proc. ACM Workshop on Standards, Interoperability and Practices, pp. 51–54 (2000)
17. Park, S.J., Park, D.K.W.C.: Core experiments on MPEG-7 edge histogram descriptor. Technical report, ISO/IEC JTC1/SC29/WG11-MPEG2000/M5984 (2000)
18. Pietikäinen, M., Nurmela, T., Mäenpää, T., Turtinen, M.: View-based recognition of real-world textures. Pattern Recognit. 37(2), 313–323 (2004)
19. Puzicha, J., Buhmann, J.M., Rubner, Y., Tomasi, C.: Empirical evaluation of dissimilarity measures for color and texture. In: Proc. International Conference on Computer Vision, vol. 2, p. 1165 (1999)
20. Sim, D.G., Kim, H.K., Oh, D.I.: Translation, scale, and rotation invariant texture descriptor for texture-based image retrieval. In: Proc. International Conference on Image Processing, vol. 3, pp. 742–745 (2000)
21. Stricker, M., Orengo, M.: Similarity of color images. In: Storage and Retrieval of Image and Video Databases III, vol. 2, pp. 381–392 (1995)
22. Swain, M., Ballard, D.: Color indexing. In: Proc. International Conference on Computer Vision, pp. 11–32 (1990)
23. Takala, V.: Local Binary Pattern Method in Context-based Image Retrieval. M.Sc. thesis, Department of Electrical and Information Engineering, University of Oulu (2004) (In Finnish)

24. Takala, V., Ahonen, T., Pietikäinen, M.: Block-based methods for image retrieval using local binary patterns. In: Scandinavian Conference on Image Analysis. Lecture Notes in Computer Science, vol. 3540, pp. 882–891. Springer, Berlin (2005)
25. Tamura, H., Mori, T., Yamawaki, T.: Textural features corresponding to visual perception. IEEE Trans. Syst. Man Cybern. **8**, 460–473 (1978)
26. Turtinen, M., Pietikäinen, M.: Contextual analysis of textured scene images. In: Proc. British Machine Vision Conference, pp. 849–858 (2006)
27. Varma, M., Zisserman, A.: Classifying images of materials: Achieving viewpoint and illumination independence. In: European Conference on Computer Vision. Lecture Notes in Computer Science, vol. 2352, pp. 255–271. Springer, Berlin (2002)
28. Varma, M., Zisserman, A.: Classifying materials from images: To cluster or not to cluster? In: Proc. International Workshop on Texture Analysis and Synthesis, pp. 139–144 (2002)
29. Yao, C.H., Chen, S.Y.: Retrieval of translated, rotated and scaled color textures. Pattern Recognit. **36**(4), 913–929 (2003)

Part III
Motion Analysis

Chapter 7
Recognition and Segmentation of Dynamic Textures

Description, recognition and segmentation of dynamic textures have attracted increasing attention. Dynamic textures provide a new and very effective tool for motion analysis. The past research on motion analysis has been based on assumption that the scene is Lambertian, rigid and static. This kind of constraints greatly limit the applicability of motion analysis [23]. When considering video sequences as dynamic textures the constraints mentioned above can be relaxed.

This chapter presents the application of spatiotemporal local binary patters (Chap. 3) to dynamic texture recognition and segmentation. Firstly, VLBP and LBP-TOP are utilized to dynamic texture recognition [25]. In two dynamic texture databases, DynTex and MIT, both descriptors clearly outperformed the earlier approaches. Rotation invariant LBP-TOP obtained promising results on recognizing dynamic textures with view variations [27]. Secondly, the problem of segmenting DT into disjoint regions in an unsupervised way [4] is addressed. Each region is characterized by histograms of local binary patterns and contrast in a spatiotemporal mode. It combines the motion and appearance of DT together. Experimental results show that this method is effective in segmenting regions that differ in their dynamics.

7.1 Dynamic Texture Recognition

7.1.1 Related Work

Dynamic texture recognition has received considerable interest in the research community. Chetverikov and Péteri [6] placed the existing approaches of temporal texture recognition into five classes: methods based on optic flow, methods computing geometric properties in the spatiotemporal domain, methods based on local spatiotemporal filtering, methods using global spatiotemporal transforms, and finally model-based methods that use estimated model parameters as features. The methods based on optic flow [10, 16, 17] are currently the most popular ones [6], because

M. Pietikäinen et al., *Computer Vision Using Local Binary Patterns*,
Computational Imaging and Vision 40,
DOI 10.1007/978-0-85729-748-8_7, © Springer-Verlag London Limited 2011

optic flow estimation is a computationally efficient and natural way to characterize the local dynamics of a temporal texture.

Key issues concerning dynamic texture recognition include: (1) combining motion features with appearance features, (2) processing locally to catch the transition information in space and time, for example the passage of burning fire changing gradually from a spark to a large fire, (3) defining features which are robust against image transformations such as rotation, (4) insensitivity to illumination variations, (5) computational simplicity, and (6) multi-resolution analysis. None of the proposed methods satisfies all these requirements. To address these issues, two kinds of spatiotemporal descriptors introduced in Chap. 3 were exploited: VLBP and LBP-TOP.

The large dynamic texture database DynTex was used to evaluate the performance of the DT recognition methods. Additional experiments with the widely used MIT dataset [17, 22] were also carried out.

7.1.2 Measures

After obtaining the local features on the basis of different parameters of L, P and R for VLBP, or P_{XY}, P_{XT}, P_{YT}, R_X, R_Y, R_T for LBP-TOP, a leave-one-group-out classification test was carried out for DT recognition based on the nearest class. If one DT includes m samples, all DT samples are separated into m groups. The performance is evaluated by letting each sample group be unknown and using the remaining $m - 1$ sample groups for training. The mean VLBP features or LBP-TOP features of all the $m - 1$ samples are computed as the feature for the class. The omitted sample is classified or verified according to its difference with respect to the class using the k nearest neighbor method ($k = 1$).

In classification, the dissimilarity between a sample and model feature distribution is measured using the log-likelihood statistic:

$$L(S, M) = -\sum_{b=1}^{B} S_b \log M_b, \tag{7.1}$$

where B is the number of bins and S_b and M_b correspond to the sample and model probabilities at bin b, respectively. Other dissimilarity measures like histogram intersection or chi square distance could also be used.

When the DT is described in the XY, XT and YT planes, it can be expected that some of the planes contain more useful information than others in terms of distinguishing between DTs. To take advantage of this, a weight can be set for each plane based on the importance of the information it contains. The weighted log-likelihood statistic is defined as:

$$L_w(S, M) = -\sum_{i,j} (w_j S_{j,i} \log M_{j,i}), \tag{7.2}$$

in which w_j is the weight of plane j.

Fig. 7.1 DynTex database

7.1.3 Multi-resolution Analysis

By altering L, P and R for VLBP, P_{XY}, P_{XT}, P_{YT}, R_X, R_Y, R_T for LBP-TOP, operators can be realized for any quantization of the time interval, the angular space and spatial resolution. Multi-resolution analysis can be accomplished by combining the information provided by multiple operators of varying (L, P, R) and $(P_{XY}, P_{XT}, P_{YT}, R_X, R_Y, R_T)$.

The most accurate information would be obtained by using the joint distribution of these codes [14]. However, such a distribution would be overwhelmingly sparse with any reasonable size of image and sequence. For example, the joint distribution of $VLBP_{1,4,1}^{riu2}$, $VLBP_{2,4,1}^{riu2}$ and $VLBP_{2,8,1}^{riu2}$ would contain $16 \times 16 \times 28 = 7168$ bins. So, only the marginal distributions of the different operators are considered, even though the statistical independence of the outputs of the different VLBP operators or simplified concatenated bins from three planes at a central pixel cannot be warranted.

In this study, a straightforward multi-resolution analysis was performed by defining the aggregate dissimilarity as the sum of individual dissimilarity between the different operators on the basis of the additivity property of the log-likelihood statistic [14]:

$$L_N = \sum_{n=1}^{N} L(S^n, M^n), \qquad (7.3)$$

where N is the number of operators and S^n and M^n correspond to the sample and model histograms extracted with operator n ($n = 1, 2, \cdots, N$).

7.1.4 Experimental Setup

The DynTex dataset (www.cwi.nl/projects/dyntex/) is a large and varied database of dynamic textures. Figure 7.1 shows example DTs from this dataset [26]. The image size is 400×300.

In the experiments on the DynTex database, each sequence was divided into eight non-overlapping subsets, but not half in X, Y and T. The segmentation position in

(a)   (b)

Fig. 7.2 (**a**) Segmentation of DT sequence. (**b**) Examples of segmentation in space

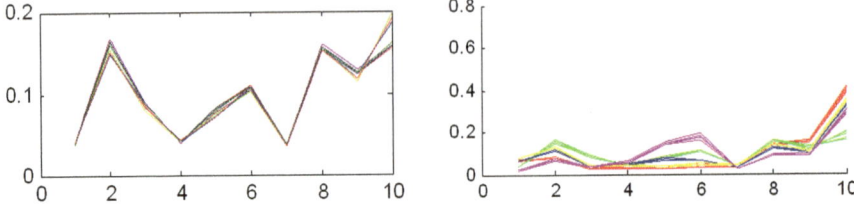

Fig. 7.3 Histograms of dynamic textures. (**a**) Histograms of up-down tide with 10 samples for VLBP$^{riu2}_{2,2,1}$. (**b**) Histograms of four classes each with 10 samples for VLBP$^{riu2}_{2,2,1}$

volume was selected randomly. For example in Fig. 7.2 [26], the transverse plane with $x = 170$, the lengthways plane with $y = 130$, in the time direction with $t = 100$ were selected. These eight samples do not overlap each other, and they have different spatial and temporal information. Sequences with the original size but only cut in the time direction are also included in the experiments. So one can get 10 samples of each class and all samples are different in image size and sequence length from each other. Figure 7.2(a) demonstrates the segmentation, and Fig. 7.2(b) shows some segmentation examples in space. It can be seen that this sampling method increases the challenge of recognition in a large database.

7.1.5 Results for VLBP

Figure 7.3(a) shows the histograms of 10 samples of a dynamic texture using VLBP$^{riu2}_{2,2,1}$. It can be seen that for different samples of the same class, their VLBP codes are very similar to each other, even if they are different in spatial and temporal variation. Figure 7.3(b) depicts histograms of 4 classes each with 10 samples as in Fig. 7.2(a). It is clear that the VLBP features have good similarity within classes and good discrimination between classes.

Table 7.1 presents the overall classification rates. For VLBP, when the number of neighboring points increases, the number of patterns for basic VLBP will become very large: 2^{3P+2}. Due to this rapid increase the feature vector will soon become too long to handle. Therefore only the results for $P = 2$ and $P = 4$ are given in Table 7.1. Using all 16384 bins of the basic VLBP$_{2,4,1}$ provides a 94.00% rate, while

Table 7.1 Results (%) in DynTex dataset. (Superscript $riu2$ means rotation invariant uniform patterns, $u2$ is the uniform patterns without rotation invariance, and ri represents the rotation invariant patterns. The numbers inside the parentheses demonstrate the length of corresponding feature vectors)

	$VLBP_{1,2,1}$	$VLBP_{2,2,1}$	$VLBP_{1,4,1}$	$VLBP_{2,4,1}$
Basic	91.71 (256)	91.43 (256)	94.00 (16384)	94.00 (16384)
$u2$	87.71 (59)	90.00 (59)	93.43 (185)	93.71 (185)
ri	89.43 (144)	90.57 (144)	93.14 (4176)	95.71 (4176)
$riu2$	83.43 (10)	83.43 (10)	88.57 (16)	85.14 (16)
Multi-resolution	$VLBP^{riu2}_{2,2,1+2,4,1}$: 90.00 ($VLBP^{riu2}_{2,2,1}$: 83.43, $VLBP^{riu2}_{2,4,1}$: 85.14)		$VLBP^{riu2}_{2,2,1+1,2,1}$: 86.00 ($VLBP^{riu2}_{1,2,1}$: 83.43, $VLBP^{riu2}_{2,2,1}$: 83.43)	

Table 7.2 Results (%) in DynTex dataset (values in square bracket are weights assigned for three sets of LBP bins)

LBP-TOP		XY	XT	YT	Con	Weighted
8, 8, 8, 1, 1, 1	$u2$	92.86	88.86	89.43	94.57	96.29 [4, 1, 1]
2, 2, 2, 1, 1, 1	Basic	70.86	60.86	78.00	88.86	90.86 [3, 1, 4]
4, 4, 4, 1, 1, 1	Basic	94.00	86.29	91.71	93.71	94.29 [6, 1, 5]
8, 8, 8, 1, 1, 1	Basic	95.14	90.86	90.00	95.43	97.14 [5, 2, 1]
8, 8, 8, 3, 3, 3	Basic	90.00	91.17	94.86	95.71	96.57 [1, 2, 5]
8, 8, 8, 3, 3, 1	Basic	89.71	91.14	92.57	94.57	95.14 [1, 2, 3]

$VLBP_{2,4,1}$ with $u2$ gives a good result of 93.71% using only 185 bins. When using rotation invariant $VLBP^{ri}_{2,4,1}$ (4176 bins), the result is 95.71%. With more complex features or multi-resolution analysis, better results could be expected. For example, the multi-resolution features $VLBP^{riu2}_{2,2,1+2,4,1}$ obtain a good rate of 90.00%, better than the results from the respective features $VLBP^{riu2}_{2,2,1}$ (83.43%) and $VLBP^{riu2}_{2,4,1}$ (85.14%). However, when using multi-resolution analysis for basic patterns, the results are not improved, partly because the feature vectors are too long.

It can be seen that the basic VLBP performs very well, but it does not allow the use of many neighboring points P. A higher value for P is shown to provide better results. With uniform patterns, the feature vector length can be reduced without much loss in performance. Rotation invariant features perform almost as well as the basic VLBP for these textures. They further reduce the feature vector length and can handle the recognition of DTs after rotation.

7.1.6 Results for LBP-TOP

Table 7.2 presents the overall classification rates for LBP-TOP. The first three columns give the results using only one histogram from the corresponding plane; they are much lower than those from direct concatenation (fourth column) and

weighted measures (fifth column). Moreover, the weighted measures of three histograms achieved better results than direct concatenation because they consider the
different contributions of features. When using only the LBP-TOP$_{8,8,8,1,1,1}$, good
results of 97.14% are obtained with the weight $[5, 2, 1]$ for the three histograms.
Because the frame rate or the resolution in the temporal axis in the DynTex is high
enough for using the same radius, it can be seen from Table 7.2 that results from a
smaller radius in the time axis T (the last row) are as good as the same radius to the
other two planes.

For selecting the weights, a heuristic approach was used which takes into account
the different contributions of the features from the three planes. First, by computing the recognition rates for each plane separately, three rates $X = [x1, x2, x3]$ are
obtained; then, it can be assumed that the lower the minimum rate, the smaller the
relative advantage will be. For example, the recognition rate improvement from 70%
to 80% is better than from 50% to 60%, even though the differences are both 10%.
The relative advantage of the two highest rates to the lowest one can now be computed as:

$$Y = (X - \min(X))/((100 - \min(X))/10). \qquad (7.4)$$

Finally, considering that the weight of the lowest rate is 1, the weights of the
other two histograms can be obtained according to a linear relationship of their
differences to that with the lowest rate. The following presents the final computing
step, and W is the generated weight vector corresponding to the three histograms:

$$Y1 = \text{round}(Y); \qquad Y2 = (Y \times ((\max(Y1) - 1))/\max(Y) + 1;$$
$$W = \text{round}(Y2). \qquad (7.5)$$

The results are very good compared to the state-of-the-art. In [11], a classification
rate of 98.1% was reported for 26 classes of the DynTex database. However, their
test and training samples were only different in the length of the sequence, but the
spatial variation was not considered. This means that their experimental setup was
much simpler. When experimenting using all 35 classes with samples having the
original image size and only different in sequence length, a 100% classification rate
using VLBP$^{u2}_{1,8,1}$ or using LBP-TOP$_{8,8,8,1,1,1}$ was obtained.

Experiments on the MIT dataset [22] were also performed using a similar experimental setup as with DynTex, obtaining a 100% accuracy both for the VLBP
and LBP-TOP. None of the earlier methods have reached the same performance, see
for example [11, 15, 17, 22]. Except for [17], which used the same segmentation
here but with only 10 classes, all other papers used simpler datasets which did not
include the variation in space and time.

Comparing VLBP and LBP-TOP, they both combine motion and appearance together, and are robust to translation and illumination variations. Their differences
are: (a) the VLBP considers the co-occurrence of all the neighboring points in three
frames of volume at the same time, while LBP-TOP only considers the three orthogonal planes making it easy to be extended to use more neighboring information;
(b) when the time interval $L > 1$, the neighboring frames with a time variance of

less than L will be missed out in VLBP, but the latter method still keeps the local information from all neighboring frames; (c) computation of the latter is much simpler with the complexity $O(XYT \cdot 2^P)$, compared to the VLBP with $O(XYT \cdot 2^{3P})$.

7.1.7 Experiments of Rotation Invariant LBP-TOP to View Variations

View variations are very common in dynamic textures. View-invariant recognition DTs is a very challenging task. Most of the proposed methods for dynamic texture categorization validated their performance on the ordinary DT databases, without viewpoint changes, except [19, 24]. Woolfe and Fitzgibbon addressed shift invariance [24] and Ravichandran et al. [19] proposed to use bag of system (BoS) as the representation for dealing with viewpoint changes.

To evaluate the robustness of the rotation invariant LBP-TOP described in Chap. 3 to view variations, the same dataset to [19, 20] was used. This dataset consists of 50 classes of four video sequences each. Many previous works [2, 20] are based on the 50 class structure and the reported results are not on the entire video sequences, but on a manually extracted patch of size 48×48 [19]. In [19], the authors combine the sequences that are taken from different viewpoints, and reduce the dataset to a nine class dataset with the classes being boiling water (8), fire (8), flowers (12), fountains (20), plants (108), sea (12), smoke (4), water (12) and waterfall (16). The numbers in parentheses represent the number of sequences in the dataset.

To compare the performance of handling view variations with the methods proposed in [20] and [19], the same experimental setup was used, i.e. (1) the plants class is left out since the number of sequences of plants far outnumbered the number of sequences for the other classes, so remaining eight classes were used in the experiments; (2) four different scenarios are explored. The first set is an easy two class problem namely the case of water vs. fountain. The second one is a more challenging two class problem namely the fountain vs. waterfall. The third set of experiments is on a four class (water, fountain, waterfall and sea) problem and the last set is on the reorganized database with eight classes. These scenarios are abbreviated as W-F (Water vs. Fountain), F-WF (Fountain vs. Waterfall), FC (Four Classes) and EC (Eight Classes). Sample frames from the video sequences in the database are shown in Fig. 7.4. For every scenario, 50% of the data is used for training and the rest for testing.

In classification, the dissimilarity between a sample and a model rotation invariant LBP-TOP distribution is measured using the $L1$ distance:

$$L_1(RI^S, RI^M) = \sum_{k=1}^{K} \left| RI^S(k) - RI^M(k) \right| \qquad (7.6)$$

in which, RI means the rotation invariant histogram and K is the number of the bins in the RI histogram.

Boiling water   Fire   Flower   Fountain   Sea   Smoke   Water   Waterfall

Fig. 7.4 Sample images from reorganized dataset from [20]

The Support Vector Machine (SVM) and the Nearest Neighbor (1NN) of $L1$ distance are used as classifiers. For SVM, the second degree polynomial kernel function is used in the experiments. Figure 7.5 shows the results using 1DHFLBP-TOP$_{8,8,1,1,1}$ and 2DHFLBP-TOP$_{8,8,1,1,1}$, and the results with term frequency (TF) and soft-weighting (SW) from [19], and DK (DS) from [20]. TF and SW are the two kinds of representation utilized in [19] to represent the videos using the code-book. DK and DS depict the methods proposed in [20], in which a single LDS is used to model the whole video sequence and the nearest neighbor and SVM classifiers are used for categorization, respectively. The first four groups in Fig. 7.5 show the results for four scenarios using SVM as classifier while the last four groups present the results using 1NN as classifier. In all four experimental setups, the second (F-WF) and last (EC) setups are considered more challenging, because in the second scenario, the viewpoints in testing are not used in the training, which makes them totally novel. In the last scenario, there are a 92 video sequences in total with varying number of sequences per class and they are from various viewpoints.

It can be seen from Fig. 7.5 that the rotation invariant LBP-TOP descriptors obtain leading accuracy for all four scenarios. Especially for more challenging fountain vs. waterfall and all eight class problems, the results for 1DHFLBP-TOP$_{8,8,1,1,1}$ with SVM are 87.5% and 86.96%, and with 1NN are 87.5% and 73.91%, respectively. These are much better than for TF (70% and 63%) and SW (76% and 80%) with SVM [19], DK (50% and 52%) and DS (56% and 47%) [19]. In addition, experiments using LBP-TOP without rotation invariant characteristics were also carried out. LBP-TOP$_{8,1}^{u2}$ obtained 75% and 71.74% with SVM for more difficult F-WF and EC scenarios, which is inferior to either of the rotation invariant descriptors. From these results and comparison on the database with real 3D viewpoint changes, it can be seen that the rotation invariant LBP-TOP descriptors can deal with view variations very effectively.

7.2 Dynamic Texture Segmentation

7.2.1 Related Work

Segmentation is one of the classical problems in computer vision [1, 13, 18]. Unsupervised static texture segmentation was presented in Chap. 4. In this section, a related approach to dynamic texture segmentation is presented. The segmentation

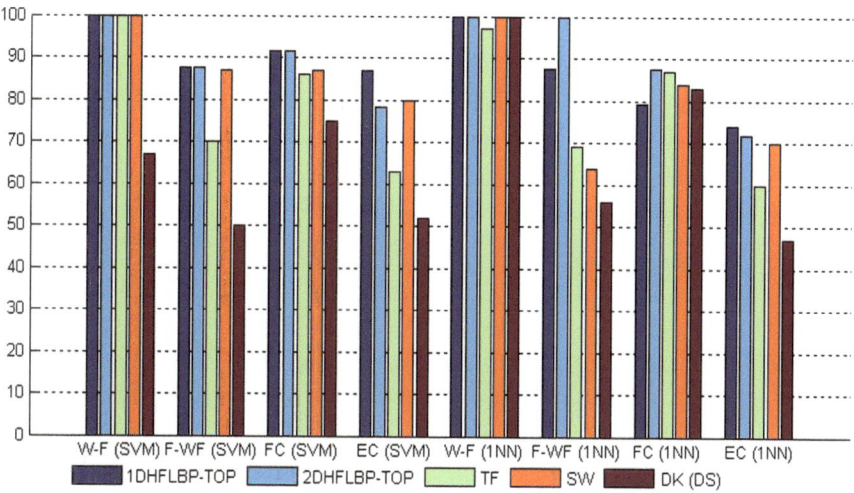

Fig. 7.5 Classification results for the four scenarios with different methods

of DTs is a challenging problem compared with the static case because of their unknown spatiotemporal extension. In general, existing approaches of DT segmentation can be generally categorized into supervised and unsupervised methods. For supervised segmentation, a priori information about the textures present is needed. In contrast, unsupervised segmentation does not need a priori information. This makes it a very challenging research problem. However, most of the recent methods need an initialization. Examples of recent approaches are methods based on mixtures of dynamic texture model [3], mixture of linear models [7], multi-phase level sets [8], Gauss-Markov models and level sets [9], Ising descriptors [12], and optical flow [23].

A key problem of DT segmentation is how to combine motion and appearance features. As discussed in Chap. 3, LBP-TOP has a promising ability to describe both the appearance and motions of DT [25]. It is also robust to monotonic gray-scale changes caused, e.g., by illumination variations.

In this section, the method for static texture segmentation introduced in Chap. 4 is extended to dynamic texture segmentation. Motivated by [25], the contrast of a single spatial texture is also generalized to a spatiotemporal mode (called as CTOP, i.e., contrast in three orthogonal planes). Combined LBP-TOP and CTOP is called as (LBP/C)TOP. It is a theoretically and computationally simple approach to model DT. (LBP/C)TOP histograms are then used for DT segmentation. The extracted features (LBP/C)TOP in a small local neighborhood reflect the spatio-temporal features of dynamic textures.

7.2.2 Features for Segmentation

After a brief review of LBP-TOP and intensity contrast, it will be described how to use the generalized $(LBP/C)_{TOP}$ for DT segmentation. More details can be found in [4].

7.2.2.1 LBP-TOP/Contrast

LBP-TOP is a spatiotemporal descriptor. As shown in Fig. 7.6, (a) is a sequence of frames (or images) of a DT; (b) denotes the three orthogonal planes or slices XY, XT and YT, where XY is the appearance (or a frame) of DT; XT shows the visual impression of a row changing in time; and YT describes the motion of a column in temporal space; (c) shows how to compute LBP and contrast for each pixel of these three planes. Here, a binary code is produced by thresholding its square neighborhood from XY, XT, YT slices independently with the value of the center pixel; (d) shows how to compute histograms by collecting up the occurrences of different binary patterns from three slices which are denoted as $H_{\lambda,\pi}$ ($\lambda = \mathrm{LBP}$ and $\pi = XY, XT, YT$). Dynamic textures are encoded by LBP using these three sub-histograms to consider simultaneously the appearance and motions in two directions, i.e., incorporating spatial domain information and two spatiotemporal co-occurrence statistics together. If these three sub-histograms $H_{\lambda,\pi}$ ($\lambda = \mathrm{LBP}$ and $\pi = XY, XT, YT$) are concatenated into a single histogram, an LBP-TOP feature histogram is obtained.

The contrast measure C is the difference between the average gray-level of those pixels which have value 1 and those which have value 0 (Fig. 7.6(c)). Likewise, the contrasts in the three orthogonal planes are also computed, which are denoted as C_π ($\pi = XY$, XT and YT). These contrast values are also regrouped in three sub-histograms $H_{\lambda,\pi}$ ($\lambda = C$ and $\pi = XY, XT, YT$). Due to the contrast C_π here refers to the features in three orthogonal planes, it is called C_{TOP}.

Let $(LBP/C)_{TOP}$ be denoted as features of LBP and C in these three planes. Thus, the $(LBP/C)_{TOP}$ feature can be denoted as a vector form:

$$\varphi = (H_{LBP,XY}, H_{\mathrm{LBP},XT}, H_{LBP,YT}, H_{C,XY}, H_{C,XT}, H_{C,YT}), \qquad (7.7)$$

where $H_{\lambda,\pi}$ ($\lambda = \mathrm{LBP}$ or C, and $\pi = XY, XT, YT$) are the six sub-histograms of LBP and contrast feature in the three orthogonal planes. Motivated by [25], the radii of $(LBP/C)_{TOP}$ in axes X, Y and T, and the number of neighboring points in XY, XT and YT planes can also be different, which are marked as R_X, R_Y and R_T, P_{XY}, P_{XT} and P_{YT}, the corresponding LBP/C feature is denoted as $(LBP/C)_{P_{XY},P_{XT},P_{YT},R_X,R_Y,R_T}^{TOP}$.

Let $l_{\mathrm{LBP}/C}$ be denoted as the size of the components of φ (i.e., $H_{\lambda,\pi}$). The "uniform patterns" are used to shorten the length of the feature vector LBP. Thus, $l_{\mathrm{LBP}} = 59$ for $H_{\mathrm{LBP},\pi}$. Meanwhile, l_C is the number of bins for $H_{C,\pi}$). According to experiments it was chosen to use $l_C = 16$ bins. See Sect. 2.5 for a detailed

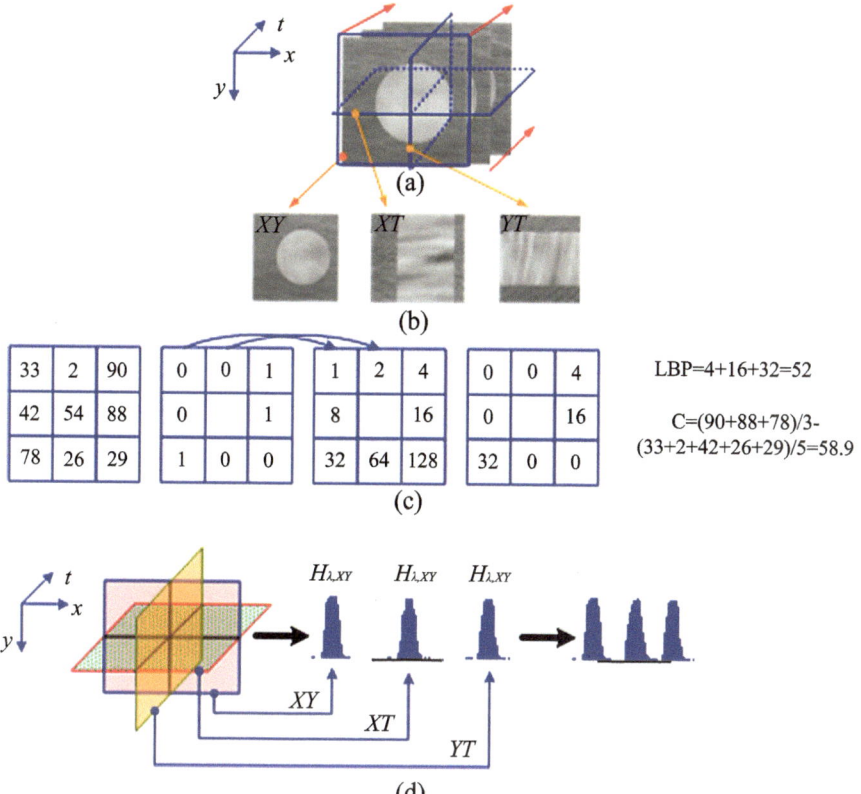

Fig. 7.6 Computation of $(LBP/C)_{TOP}$ for a DT. (**a**) A sequence of images of a DT; (**b**) Three orthogonal planes of $(LBP/C)_{TOP}$; (**c**) Computation of LBP and contrast measure C; (**d**) Computation of sub-histograms for $(LBP/C)_{TOP}$

description of the mapping from the continuous C space to the discrete bin index. In addition, it was also experientially set $R_X = 1$, $R_Y = 1$ and $R_T = 3$, $P_{XY} = 8$, $P_{XT} = 8$ and $P_{YT} = 8$. Thus, $(LBP/C)_{TOP}$ is denoted as $(LBP/C)_{8,8,8,1,1,3}^{TOP}$.

7.2.2.2 LBP-TOP/Contrast for Segmentation

Next, the measure of histogram similarity for the feature $(LBP/C)_{TOP}$ and its use it for the segmentation of DTi is described.

To compute the distance between two given histograms H_1 and H_2, the histogram intersection $\Pi(H_1, H_2)$ is used as a similarity measurement of two normalized histograms:

$$\Pi(H_1, H_2) = \sum_{i=1}^{L} \min(H_{1,i}, H_{2,i}), \qquad (7.8)$$

where L is the number of bins in a histogram.

Using the $(LBP/C)_{TOP}$ feature vector φ shown in Eq. 7.7, the distance between any two sub-blocks can also be represented as a column vector:

$$d = (\Pi_{\text{LBP},XY}, \Pi_{\text{LBP},XT}, \Pi_{\text{LBP},YT}, \Pi_{C,XY}, \Pi_{C,XT}, \Pi_{C,YT})^T. \qquad (7.9)$$

Each entry of d is computed by Eq. 7.8 using the corresponding component of φ (i.e., $H_{\lambda,\pi}$).

Let Π_λ ($\lambda = $ LBP or C) be denoted as the similarity measure of LBP_{TOP} or C_{TOP} between any two sub-blocks. It is then computed as:

$$\Pi_\lambda = \omega_{XY}\Pi_{\lambda,XY} + \omega_{XT}\Pi_{\lambda,XT} + \omega_{YT}\Pi_{\lambda,YT}, \qquad (7.10)$$

where ω_π ($\pi = XY, XT, YT$) are the weights for each plane since it was found that these three planes do not play the same role for the representation of DT.

Thus, the similarity measurement between two sub-blocks is computed as:

$$\Pi = \omega_{\text{LBP}} \times \Pi_{\text{LBP}} + \omega_C \times \Pi_C, \qquad (7.11)$$

where ω_{LBP} and ω_C are two weights to balance the effects of the LBP and contrast features. The values of the weights were experientially set as $\omega_{\text{LBP}} = 0.4$, $\omega_C = 0.6$, and $\omega_{XY} = 0.2$, $\omega_{XTY} = 0.4$, $\omega_{YT} = 0.4$.

7.2.3 Segmentation Procedure

This subsection describes how to use $(LBP/C)_{TOP}$ for segmentation. As shown in Fig. 7.7, the segmentation method consists of three phases: hierarchical splitting, agglomerative merging and pixelwise classification. Firstly, each frame is split into regions of roughly uniform texture. Then, similar adjacent regions are merged, and finally a pixelwise classification is performed to improve the boundary localization.

7.2.3.1 Splitting

Each input frame is recursively split into square blocks of varying size. The decision whether a block is split into four sub-blocks is based on a uniformity test. However, a necessary prerequisite for the following merging to be successful is that the individual frame regions are uniform in texture. To this task, the following criterion is used to perform the splitting: if *one* of the features in the three planes (i.e., LBP_π and C_π, $\pi = XY, XT, YT$) votes for splitting of current block, the splitting is performed.

Formally, the six pairwise distances D between $(LBP/C)_{TOP}$ histograms from four sub-blocks are measured as shown in Eq. 7.9 (i.e., $D = d_0, d_1, \cdots, d_5$). For

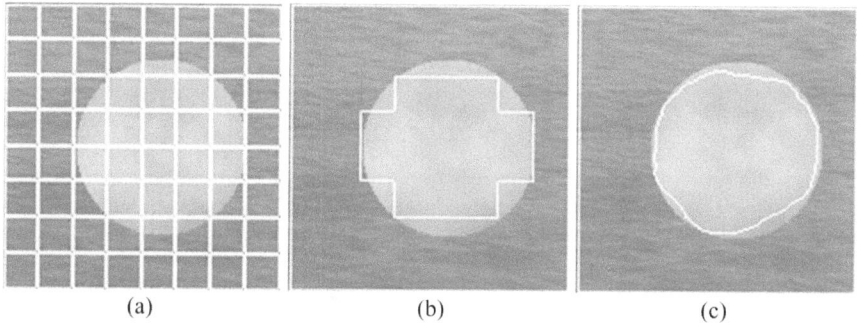

Fig. 7.7 Illustration of using $(LBP/C)_{TOP}$ for DT segmentation; (**a**) splitting, (**b**) merging, and (**c**) pixelwise classification

the distance matrix D, its largest and smallest entries of each column are computed and denoted as $D_{j\text{-}Max}$ and $D_{j\text{-}Min}$ ($j = 0, 1, \cdots, 5$), respectively. Thus, the ratio between them is:

$$R_j = D_{j\text{-}Max}/D_{j\text{-}Min} \quad (j = 0, 1, \cdots, 5). \tag{7.12}$$

The current block is split into four sub-blocks if one of the measures R_j within this block is greater than a threshold X (its value was experimentally set as 1.1):

$$R_j > X \quad (j = 0, 1, \cdots, 5). \tag{7.13}$$

This procedure is repeated recursively on each sub-block until a predetermined minimum block size S_{min} is reached. Here, a value of $S_{min} = 16$ was used.

7.2.3.2 Merging

After the input frame has been split into blocks of roughly uniform texture, similar adjacent regions are merged until a stopping criterion is satisfied. For each merging, that pair of adjacent segments is merged, which has the smallest merger importance (MI) value. Here, MI is defined as:

$$MI = f(p) \times (1 - \Pi), \tag{7.14}$$

where Π is the $(LBP/C)_{TOP}$ histogram similarity between two regions computed as shown in Eq. 7.11. The variable p is the percentage of pixels in the smaller one of the two regions. It is computed as:

$$p = N_b/N_f, \tag{7.15}$$

where N_b is the number of pixels in current block, and N_f is the number of pixels in current frame. In addition:

$$f(p) = \text{sigmoid}(\beta p) \quad (\beta = 1, 2, 3, \cdots). \tag{7.16}$$

Before moving to the next merger the distances between the new region and its all adjacent regions are also computed as shown in Eq. 7.11. The stopping rule for merging is computed as:

$$MIR = MI_{cur}/MI_{max}. \tag{7.17}$$

Here, the ratio of MI_{cur} is the merger importance for the current best merger, and MI_{max} is the largest merger importance of all preceding mergers. Thus, merging is stopped if MIR is larger than a preset threshold Y (we experimentally set its value as 2.0):

$$MIR > Y. \tag{7.18}$$

7.2.3.3 Pixelwise Classification

After the splitting and merging process, a simple pixelwise classification is performed to improve the localization of the boundaries. To this end, it is switched into a texture classification mode by using $(LBP/C)_{TOP}$ histograms of the frame segments as the texture models. For each boundary pixel (i.e., at least one of its 4-connected neighbors with a different label), $(LBP/C)_{TOP}$ feature histograms (as shown in Fig. 7.6) over its circular neighbor $B(r)$ are computed, where r is the radius of the circular neighbor. Then the similarity between the neighbor histograms and the models of those regions (which are 4-connected to the pixel in query) is computed. The similarity computation is as mentioned in Eq. 7.11. The pixel is relabeled if the label of the nearest model votes a different label from the current label of the pixel. In experiments, the value of 11 was used for radius r.

In the next scan over the frame only the neighborhoods of those pixels are checked, which were relabeled in the previous sweep. The process of pixelwise classification continues until no pixels are relabeled or maximum number of sweeps is reached. This is set to be the same value of radius r.

7.2.4 Experiments

Figure 7.8 presents some experiments performed on various types of sequences [4]. Here, Fig. 7.8(e) shows the results of the presented method and Fig. 7.8(a) the results using the traditional LBP and contrast features to segment each frame without using dynamic information provided by its neighbor frames. Figure 7.8(b) depicts the results using LBP-TOP to segment DT but not using the contrast features. One can conclude that both the dynamic information (provided by the two planes XT, YT as shown in Fig. 7.6(b)) and contrast are valuable for the segmentation of DT.

In addition, in Fig. 7.8(c) and (d), the results for [9] and [12] (unsupervised methods with initialization) are also given. An important improvement of presented method over the method in [9] is that one can handle regions with moving boundaries. Compared to [9] and [12], the proposed method is an unsupervised method without any initialization.

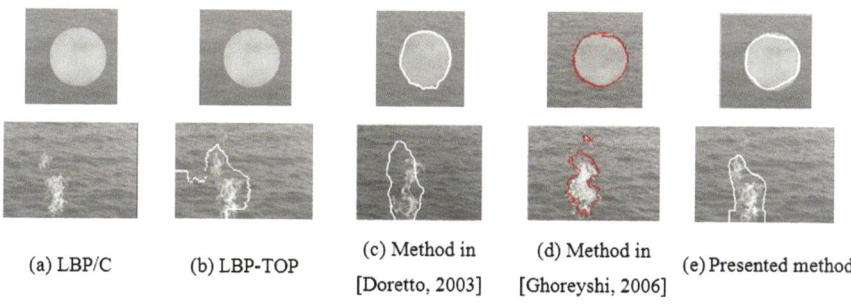

| (a) LBP/C | (b) LBP-TOP | (c) Method in [Doretto, 2003] | (d) Method in [Ghoreyshi, 2006] | (e) Presented method |

Fig. 7.8 Some experimental results performed on various types of sequences and compared with existing methods

| (a) Frame 13 | (b) Frame 19 |

Fig. 7.9 Some experimental results performed on a real challenging sequence (the sequence is from [21])

Finally, Fig. 7.9 presents results on a real sequence taken from [21] (two men in the swing) [4]. From this figure (all the parameters are the same to Fig. 7.8 except $Y = 1.6$), the performance of the algorithm on tracking the boundary of the two people in the swing throughout the sequence is shown. One can find that the algorithm performs fairly well on this challenging sequence with clustered background.

7.3 Discussion

This chapter presented the applications of spatiotemporal LBP to dynamic texture recognition and segmentation.

An approach to dynamic texture recognition was exploited, in which volume LBP and LBP-TOP operators are used to combine the motion and appearance together. Experiments on two databases with a comparison to the state-of-the-art results showed that this method is efficient for DT recognition. The approach is robust in terms of grayscale and rotation variations, making it very promising for real application problems. Experiments using rotation invariant LBP-TOP on dynamic textures captured from different views provided better results than the state-of-the-art, proving the effectiveness of this approach for dealing with view variations.

The problem of segmenting DT into disjoint regions in an unsupervised way was considered. Each region is characterized by histograms of local binary patterns and contrast in a spatiotemporal mode, combining the motion and appearance of DT together. Experimental results and a comparison to some existing methods showed that the presented method is quite effective for DT segmentation, and is also computationally simple compared to methods such as those using mixtures of dynamic texture model or level sets. Furthermore, the method performs fairly well on a sequence with clustered background. Recently, Chen et al. [5] proposed to combine LBP-TOP with a differential excitation measure of Weber Law Descriptor (WLD) and use statistical learning to determine the thresholds for further improving the DT segmentation performance.

References

1. Amiaz, T., Fazekas, S., Chetverikov, D., Kiryati, N.: Detecting regions of dynamic texture. In: First International Conference on Scale Space and Variational Methods in Computer Vision, pp. 848–859 (2007)
2. Chan, A., Vasconcelos, N.: Classifying video with kernel dynamic textures. Proc. IEEE Conference on Computer Vision and Pattern Recognition, pp. 1–6 (2007)
3. Chan, A.B., Vasconcelos, N.: Modeling, clustering, and segmenting video with mixtures of dynamic texture. IEEE Trans. Pattern Anal. Mach. Intell. **30**(5), 909–926 (2008)
4. Chen, J., Zhao, G., Pietikäinen, M.: Unsupervised dynamic texture segmentation using local spatiotemporal descriptors. In: Proc. International Conference on Pattern Recognition, pp. 1–4 (2008)
5. Chen, J., Zhao, G., Pietikäinen, M.: An improved local descriptor and threshold learning for unsupervised dynamic texture segmentation. In: Proc. ICCV Workshop on Machine Learning for Vision-Based Motion Analysis, pp. 460–467 (2009)
6. Chetverikov, D., Péteri, R.: A brief survey of dynamic texture description and recognition. In: Proc. International Conference on Computer Recognition Systems, pp. 17–26 (2005)
7. Cooper, L., Liu, J., Huang, K.: Spatial segmentation of temporal texture using mixture linear models. In: Proc. ICCV Workshop on Dynamical Vision (2005)
8. Cremers, D., Soatto, S.: Motion competition: A variational approach to piecewise parametric motion segmentation. Int. J. Comput. Vis. **62**(3), 249–265 (2005)
9. Doretto, G., Chiuso, A., Wu, Y.N., Soatto, S.: Dynamic texture segmentation. In: Proc. International Conference on Computer Vision, pp. 1236–1242 (2003)
10. Fablet, R., Bouthemy, P.: Motion recognition using nonparametric image motion models estimated from temporal and multiscale co-occurrence statistics. IEEE Trans. Pattern Anal. Mach. Intell. **25**, 1619–1624 (2003)
11. Fazekas, S., Chetverikov, D.: Normal versus complete flow in dynamic texture recognition: A comparative study. In: Proc. International Workshop on Texture Analysis and Synthesis, pp. 37–42 (2005)
12. Ghoreyshi, A., Vidal, R.: Segmenting dynamic textures with Ising descriptors, ARX models and level sets. In: Dynamical Vision. Lecture Notes in Computer Science, vol. 4358, pp. 127–141 (2007)
13. Murray, D.W., Buxton, B.F.: Scene segmentation from visual motion using global optimization. IEEE Trans. Pattern Anal. Mach. Intell. **9**(2), 220–228 (1987)

14. Ojala, T., Pietikäinen, M., Mäenpää, T.: Multiresolution gray-scale and rotation invariant texture classification with local binary patterns. IEEE Trans. Pattern Anal. Mach. Intell. **24**(7), 971–987 (2002)
15. Otsuka, K., Horikoshi, T., Suzuki, S., Fujii, M.: Feature extraction of temporal texture based on spatiotemporal motion trajectory. In: Proc. International Conference on Pattern Recognition, vol. 2, pp. 1047–1051 (1998)
16. Peh, C.H., Cheong, L.-F.: Synergizing spatial and temporal texture. IEEE Trans. Image Process. **11**, 1179–1191 (2002)
17. Péteri, R., Chetverikov, D.: Dynamic texture recognition using normal flow and texture regularity. In: Proc. Iberian Conference on Pattern Recognition and Image Analysis, pp. 223–230 (2005)
18. Polana, R., Nelson, R.: Temporal texture and activity recognition. In: Motion-Based Recognition, pp. 87–115 (1997)
19. Ravichandran, A., Chaudhry, R., Vidal, R.: View-invariant dynamic texture recognition using a bag of dynamical systems. In: IEEE Conference on Computer Vision and Pattern Recognition, pp. 1–6 (2009)
20. Saisan, P., Doretto, G., Wu, Y.N., Soatto, S.: Dynamic texture recognition. In: Proc. IEEE Conference on Computer Vision and Pattern Recognition, vol. 2, pp. 58–63 (2001)
21. Schodl, A., Essa, I.: Controlled animation of video sprites. In: Proc. ACM SIGGRAPH, 2002
22. Szummer, M., Picard, R.W.: Temporal texture modeling. In: Proc. IEEE International Conference on Image Processing, vol. 3, pp. 823–826 (1996)
23. Vidal, R., Ravichandran, A.: Optical flow estimation and segmentation of multiple moving dynamic textures. In: Proc. IEEE Conference on Computer Vision and Pattern Recognition, vol. 2, pp. 516–521 (2005)
24. Woolfe, F., Fitzgibbon, A.: Shift-invariant dynamic texture recognition. In: Proc. European Conference on Computer Vision, pp. 549–562 (2006)
25. Zhao, G., Pietikäinen, M.: Dynamic texture recognition using local binary patterns with an application to facial expressions. IEEE Trans. Pattern Anal. Mach. Intell. **29**(6), 915–928 (2007)
26. Zhao, G., Pietikäinen, M.: Dynamic texture recognition using volume local binary patterns. In: Dynamical Vision. Lecture Notes in Computer Science, vol. 4358, pp. 165–177. Springer, Berlin (2007)
27. Zhao, G., Ahonen, T., Matas, J., Pietikäinen, M.: Rotation invariant image and video description with local binary pattern features. Under review (2011)

Chapter 8
Background Subtraction

Background subtraction is often among the first tasks in computer vision applications, making it a critical part of the system. The output of background subtraction is an input to a higher-level process that can be, for example, the tracking of an identified object. The performance of background subtraction depends mainly on the background modeling technique used. Especially natural scenes set many challenges for background modeling since they are usually dynamic, including illumination changes, swaying vegetation, rippling water, flickering monitors etc. A robust background modeling algorithm should also be able to handle situations where new objects are introduced to or old ones removed from the background. Furthermore, the shadows of the moving and scene objects can cause problems. Even in a static scene frame-to-frame changes can occur due to noise and camera jitter. Moreover, the background modeling algorithm should operate in real-time.

In this chapter, an approach that uses discriminative texture features to capture background statistics is presented [3]. An early version of the method based on block-wise processing was introduced in [4]. The presented method tries to address all the issues mentioned above except the handling of shadows which has turned out to be an extremely difficult problem to solve with background modeling.

8.1 Related Work

A wide variety of different methods for detecting moving objects have been proposed and many different features are utilized for modeling the background [3].

A very popular way is to model each pixel in a video frame with a Gaussian distribution. This is the underlying model for many background subtraction algorithms. A simple technique is to calculate an average image of the scene, to subtract each new video frame from it and to threshold the result. The adaptive version of this algorithm updates the model parameters recursively by using a simple adaptive filter. This single Gaussian model was used in [12]. The previous model does not work well in the case of dynamic natural environments since they include repetitive motions like swaying vegetation, rippling water, flickering monitors, camera jitter etc.

M. Pietikäinen et al., *Computer Vision Using Local Binary Patterns*,
Computational Imaging and Vision 40,
DOI 10.1007/978-0-85729-748-8_8, © Springer-Verlag London Limited 2011

This means that the scene background is not completely static. By using more than one Gaussian distribution per pixel it is possible to handle such backgrounds. In [2], the mixture of Gaussians approach was used in a traffic monitoring application. The model for pixel intensity consisted of three Gaussian distributions corresponding to the road, vehicle and shadow distributions. One of the most commonly used approaches for updating the Gaussian mixture model was presented in [9]. Instead of using the exact EM algorithm, an online K-means approximation was used. Many authors have proposed improvements and extensions to this algorithm.

The Gaussian assumption for the pixel intensity distribution does not always hold. To deal with the limitations of parametric methods, a non-parametric approach to background modeling was proposed in [1]. The proposed method utilizes a general non-parametric kernel density estimation technique for building a statistical representation of the scene background. The probability density function for pixel intensity is estimated directly from the data without any assumptions about the underlying distributions. In [6], a quantization/clustering technique to construct a non-parametric background model was presented. The background is encoded on a pixel by pixel basis and samples at each pixel are clustered into the set of codewords.

The handling of shadows has turned out to be an extremely difficult problem to solve with background modeling. In [8], a comprehensive survey of moving shadow detection approaches is presented.

8.2 An LBP-based Approach

Due to its robustness against illumination changes and its computational simplicity, the LBP is very suitable for background modeling. Differing from most other approaches, the features in background modeling are computed over a larger area than a single pixel. In this section, this approach is described in more detail. The algorithm can be divided into two phases, *background modeling* and *foreground detection*, described in Sects. 8.2.2 and 8.2.3.

8.2.1 Modifications of the LBP Operator

The basic LBP has a limitation that it does not work very robustly on flat image areas such as sky, where the gray values of the neighboring pixels are very close to the value of the center pixel. This is due to the thresholding scheme of the operator. In order to make the LBP more robust against these negligible changes in pixel values, the thresholding scheme of the operator was modified by replacing the term $s(g_p - g_c)$ in Eq. 2.10 with the term $s(g_p - g_c + a)$. The bigger the value of $|a|$ is, the bigger changes in pixel values are allowed without affecting the thresholding results. In order to retain the discriminative power of the LBP operator, a relatively small value should be used. In the experiments a a value of 3 was used. The results presented in [4] show that good results can also be achieved by using the original LBP. With the modified version, the presented background subtraction method consistently behaves more robustly and thus should be preferred over the original one.

8.2.2 Background Modeling

Background modeling is the most important part of any background subtraction algorithm. The goal is to construct and maintain a statistical representation of the scene that the camera sees. As in most earlier studies, the camera is assumed to be non-moving. Each pixel of the background is modeled identically, which allows a high speed parallel implementation if needed. Texture information is utilized in modeling the background and the LBP was selected as the measure of texture because of its good properties. In the following, the background model update procedure for one pixel is explained, but the procedure is identical for each pixel.

The feature vectors of a particular pixel over time are considered as a pixel process. The LBP histogram computed over a circular region of radius R_{region} around the pixel is used as the feature vector. The radius R_{region} is a user-settable parameter. The background model for the pixel consists of a group of adaptive LBP histograms, $\{M_0, \ldots, M_{K-1}\}$, where K is selected by the user. Each model histogram has a weight between 0 and 1 so that the weights of the K model histograms sum up to one. The weight of the kth model histogram is denoted by ω_k.

Let the LBP histogram of the given pixel computed from the new video frame be denoted by H. At the first stage of processing, H is compared to the current K model histograms using a proximity measure. The histogram intersection is used as the measure in the experiments:

$$H(A, B) = \sum_{n=0}^{N-1} \min(A_n, B_n), \tag{8.1}$$

where A and B are the histograms and N is the number of histogram bins. This measure has an intuitive motivation in that it calculates the common part of two histograms. Its advantage is that it explicitly neglects features which only occur in one of the histograms. The complexity is very low as it requires very simple operations only. The complexity is linear for the number of histogram bins: $O(N)$. It is also possible to use other measures such as chi square. The threshold for the proximity measure, T_P, is a user-settable parameter.

If the proximity is below the threshold T_P for all model histograms, the model histogram with the lowest weight is replaced with H. The new histogram is given a low initial weight. In the experiments, a value of 0.01 was used. No further processing is required in this case.

More processing is required if matches were found. The best match is selected as the model histogram with the highest proximity value. The best matching model histogram is adapted with the new data by updating its bins as follows:

$$M_k = \alpha_b H + (1 - \alpha_b) M_k, \quad \alpha_b \in [0, 1], \tag{8.2}$$

where α_b is a user-settable learning rate. Furthermore, the weights of the model histograms are updated:

$$\omega_k = \alpha_w M_k + (1 - \alpha_w) \omega_k, \quad \alpha_w \in [0, 1], \tag{8.3}$$

where α_w is another user-settable learning rate and M_k is 1 for the best matching histogram and 0 for the others. The adaptation speed of the background model is controlled by the learning rate parameters α_b and α_w. The bigger the learning rate the faster the adaptation.

All of the model histograms are not necessarily produced by the background processes. The persistence of the histogram can be used in the model to decide whether the histogram models the background or not. It can be seen from Eq. 8.3 that the persistence is directly related to the histogram's weight: the bigger the weight, the higher the probability of being a background histogram. As a last stage of the updating procedure, the model histograms are sorted in decreasing order according to their weights, and the first B histograms are selected as the background histograms:

$$\omega_0 + \cdots + \omega_{B-1} > T_B, \quad T_B \in [0, 1], \tag{8.4}$$

where T_B is a user-settable threshold.

8.2.3 Foreground Detection

Foreground detection is performed before updating the background model. The histogram H is compared to the current B background histograms using the same proximity measure as in the update algorithm. If the proximity is higher than the threshold T_P for at least one background histogram, the pixel is classified as background. Otherwise, the pixel is marked as foreground.

8.3 Experiments

Figure 8.1 shows results for the method using some test sequences from [3, 4]. The first two frames on the upper left are from an indoor sequence where a person is walking in a laboratory room. Background subtraction methods utilizing only color information will probably fail to detect the moving object correctly due to the similar color of the foreground and the background. The next two frames are from an indoor sequence where a person walks toward the camera. Many adaptive pixel-based methods would output a large amount of false negatives on the inner areas of the moving object, because the pixel values stay almost the same over time. The

Fig. 8.1 Some detection results of the presented method. The *first row* contains the original video frames. The *second row* contains the corresponding processed frames. The image resolution is 320 × 240 pixels. (From M. Heikkilä and M. Pietikäinen, A texture-based method for modeling the background and detecting moving objects, IEEE Transactions on Pattern Analysis and Machine Intelligence, Vol. 28, Num. 4, 657–662, 2006. @2006 IEEE)

presented method gives good results, because it utilizes information gathered over a larger area than a single pixel. The first two frames on the lower left are from an outdoor sequence which contains relatively small moving objects. The original sequence has been taken from the PETS database (ftp://pets.rdg.ac.uk/). The presented method successfully handles this situation and all the moving objects are detected correctly. The last two frames are from an outdoor sequence that contains heavily swaying trees and rippling water. This is a very difficult scene for background modeling methods. Since the approach was designed to handle also multimodal backgrounds, it manages the situation relatively well. The values for the parameters of the method are given in Table 8.1. For the first three sequences, the values remained the same. For the last sequence, values for the parameters K and T_B were changed to adjust the method for increased multimodality of the background. Some guidelines for parameter selection are given in [3].

In [11], a test set for evaluating background subtraction methods was presented. It consists of seven video sequences, each addressing a specific canonical background subtraction problem. In the same paper, ten different methods were compared using the test set. The presented method was tested against this test set, achieving the results shown in Fig. 8.2 [3]. When compared to the results in [11], the overall perfor-

Table 8.1 The parameter values of the method for the results in Figs. 8.1 and 8.2

Fig.	Sequence(s)	$LBP_{P,R}$	R_{region}	K	T_B	T_P	α_b	α_w
8.1	1–3	$LBP_{6,2}$	9	3	0.4	0.65	0.01	0.01
8.1	4	$LBP_{6,2}$	9	5	0.8	0.65	0.01	0.01
8.2	1–7	$LBP_{6,2}$	9	4	0.8	0.70	0.01	0.01

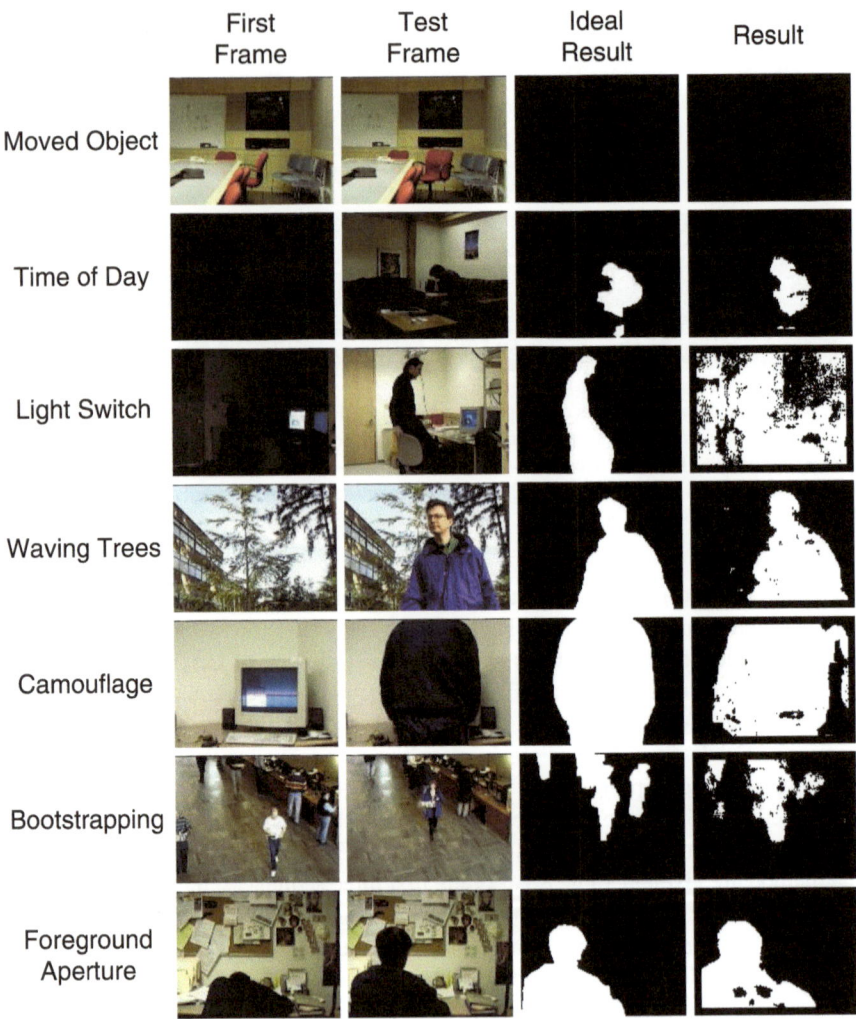

Fig. 8.2 Detection results of the method for the test sequences presented in [11]. The image resolution is 160×120 pixels. (From M. Heikkilä and M. Pietikäinen, A texture-based method for modeling the background and detecting moving objects, IEEE Transactions on Pattern Analysis and Machine Intelligence, Vol. 28, Num. 4, 657–662, 2006. @2006 IEEE)

mance of the LBP-based method seems to be better than that of the other methods. The parameter values of the method were not changed between the test sequences, although better results could be obtained by customizing the values for each sequence. See Table 8.1 for the parameter values used. Like most of the other methods, the LBP-based method was not capable of handling the *Light Switch* problem. This is because no higher level processing was utilized to detect sudden changes in background.

For the parameter values used in the tests, a frame rate of 15 fps was achieved. A standard PC with a 1.8 GHz processor and 512 MB of memory were used in experiments. The image resolution was 160×120 pixels. This makes the method well suited to systems that require real-time processing.

8.4 Discussion

A texture-based approach to background subtraction was presented, in which the background is modeled using texture features. The features are extracted by using the modified local binary pattern (LBP) operator. The approach provides with several advantages compared to earlier methods. Due to the invariance of the LBP features with respect to monotonic gray-scale changes, the method can tolerate considerable illumination variations common in natural scenes. Unlike many other approaches, the features used are very fast to compute, which is an important property from the practical implementation point of view. The presented method belongs to non-parametric methods, which means that no assumptions about the underlying distributions are needed.

The method has been evaluated against several video sequences including both indoor and outdoor scenes. It has proved to be tolerant to illumination variations, the multimodality of the background, and the introduction/removal of background objects. Furthermore the method is capable of real-time processing. Comparisons to other approaches presented in the literature have shown that the approach is very powerful when compared to the state-of-the-art.

Takala and Pietikäinen used this method successfully as the first step in their multi-object tracking method [10]. In their tracking algorithm color, texture and motion features were used to find moving objects in subsequent frames. The LBP was used as the texture feature. The LBP-based background subtraction approach has also been a basis for some new background subtraction methods. For example, Yao and Odobez proposed a multi-layer background subtraction based on color and texture [13]. LBP features work robustly on rich texture regions, whereas color features with an illumination invariant model provide more stable results in uniform regions. Good results were reported in a large variety of test cases. Hu et al. presented an integrated background model based on primal sketch and 3D scene geometry [5]. The background is divided into flat, sketchable and textured regions, and they are modeled by mixtures of Gaussians, image primitives and LBP histograms, respectively. Additionally, 3D geometry context was introduced to improve the background model. Liao et al. [7] extended the LBP operator to a scale invariant local ternary pattern (SILTP) operator (see Sect. 2.9.3), which is effective for handling illumination variations, especially soft cast shadows. Excellent results were reported for different types of video sequences.

References

1. Elgammal, A.M., Duraiswami, R., Harwood, D., Davis, L.S.: Background and foreground modeling using nonparametric kernel density estimation for visual surveillance. Proc. IEEE **90**, 1151–1163 (2002)
2. Friedman, N., Russell, S.J.: Image segmentation in video sequences: A probabilistic approach. In: Proc. Conference on Uncertainty in Artificial Intelligence, pp. 175–181 (1997)
3. Heikkilä, M., Pietikäinen, M.: A texture-based method for modeling the background and detecting moving objects. IEEE Trans. Pattern Anal. Mach. Intell. **28**(4), 657–662 (2006)
4. Heikkilä, M., Pietikäinen, M., Heikkilä, J.: A texture-based method for detecting moving objects. In: Proc. British Machine Vision Conference, pp. 187–196 (2004)
5. Hu, W., Gong, H., Zhu, S.-C., Wang, Y.: An integrated background model for video surveillance based on primal sketch and 3D scene geometry. In: Proc. IEEE Conference on Computer Vision and Pattern Recognition, pp. 1–8 (2008)
6. Kim, K., Chalidabhongse, T.H., Harwood, D., Davis, L.S.: Background modeling and subtraction by codebook construction. In: Proc. IEEE International Conference on Image Processing, pp. 3061–3064 (2004)
7. Liao, S., Zhao, G., Kellokumpu, V., Pietikäinen, M., Li, S.Z.: Modeling pixel process with scale invariant local patterns for background subtraction in complex scenes. In: Proc. IEEE Conference on Computer Vision and Pattern Recognition, p. 8 (2010)
8. Prati, A., Mikic, I., Trivedi, M.M., Cucchiara, R.: Detecting moving shadows: Algorithms and evaluation. IEEE Trans. Pattern Anal. Mach. Intell. **25**, 918–923 (2003)
9. Stauffer, C., Grimson, W.E.L.: Adaptive background mixture models for real-time tracking. In: Proc. IEEE Conference Computer Vision and Pattern Recognition, vol. 2, pp. 246–252 (1999)
10. Takala, V., Pietikäinen, M.: Multi-object tracking using color, texture and motion. In: Proc. IEEE International Workshop on Visual Surveillance, pp. 1–7 (2007)
11. Toyama, K., Krumm, J., Brumitt, B., Meyers, B.: Wallflower: Principles and practice of background maintenance. In: Proc. International Conference on Computer Vision, vol. 1, pp. 255–261 (1999)
12. Wren, C.R., Azarbayejani, A., Darrell, T., Pentland, A.P.: Pfinder: Real-time tracking of the human body. IEEE Trans. Pattern Anal. Mach. Intell. **19**(7), 780–785 (1997)
13. Yao, J., Odobez, J.-M.: Multi-layer background subtraction based on color and texture. In: Proc. IEEE Conference on Computer Vision and Pattern Recognition, pp. 1–8 (2008)

Chapter 9
Recognition of Actions

Human action recognition has become an important research topic in computer vision in recent years. It has gained a lot of attention because of its important application domains like video indexing, surveillance, human computer interaction, video games, sport video analysis, intelligent environments etc. All these application domains do have their own demands, but in general, algorithms must be able to detect and recognize various actions in real time. Also as people look different and move differently, the designed algorithms must be able to handle variations in performing actions and handle various kinds of environments.

In this chapter, the ideas of spatiotemporal analysis and the use of local features are utilized for motion description [13]. Two methods are introduced. The first one uses temporal templates to capture movement dynamics and then uses texture features to characterize the observed movements [11]. This idea is then extended into a spatiotemporal space and human movements are described with dynamic texture features [10]. Following recent trends in computer vision, the method is designed to work with image data rather than silhouettes. The performance of these methods is verified on the popular Weizmann and KTH datasets, achieving high accuracy.

9.1 Related Work

Many approaches for human activity recognition have been proposed in the literature [5, 20]. Recently there has been a lot of attention towards analyzing human motions in spatiotemporal space instead of analyzing each frame of the data separately.

The first steps to spatiotemporal analysis were taken by Bobick and Davis [2]. They used Motion Energy Images (MEI) and Motion History Images (MHI) as temporal templates to recognize aerobics movements. Matching was done using seven Hu moments. 3D extension of the temporal templates was proposed by Weinland et al. [30]. They used multiple cameras to build motion history volumes and performed action classification using Fourier analysis in cylindrical coordinates. Related 3D approaches have been used by Blank et al. [1] and Yilmaz and Shah [33] who utilized time as the third dimension and built spacetime volumes in (x, y, t) space. Space time volumes were matched using features from Poisson equations and geometric surface properties, respectively. Ke et al. [8] built a cascade of filters based on volumetric features to detect and recognize human actions. Shechtman and Irani [27] used a correlation based method in 3D whereas Kobyashi and Otsu [16] used Cubic Higher-order Local Autocorrelation to describe human movements.

Interest point based methods that have been quite popular in object recognition have also found their way to actions recognition. Laptev and Lindeberg [17] extended the Harris detector into space time interest points and detected local structures that have significant local variation in both space and time. The representation was later applied to human action recognition using Support Vector Machine (SVM) [24]. Dollár et al. [4] described interest points with cuboids, whereas Niebles and Fei-Fei [21] used a collection of spatial and spatial temporal features extracted in static and dynamic interest points.

9.2 Static Texture Based Description of Movements

Temporal templates, MEI and MHI, were introduced to describe motion information from images by [2]. MHI is a grayscale image that describes how the motion has occurred by showing more recent movements with brighter values. MEI on the other hand is a binary image that describes where the motion has occurred.

The static texture based method uses the temporal templates as a preprocessing stage to build a representation of motion. Motion is then characterized with a texture based description by extracting spatially enhanced LBP histograms from the templates.

In the presented method, silhouettes are used as input for the system. MHI can be calculated from the silhouette representation as

$$\text{MHI}_\tau(x, y, t) = \begin{cases} \tau, & \text{if } D(x, y, t) = 1; \\ \max(0, \text{MHI}_\tau(x, y, t-1) - 1), & \text{otherwise,} \end{cases} \quad (9.1)$$

where D is the absolute of silhouette difference between frames t and $t-1$, i.e., $|S(t) - S(t-1)|$. The MEI, on the other hand, can be calculated directly from the silhouettes S:

$$\text{MEI}_\tau(x, y, t) = \bigcup_{i=0}^{\tau} S(x, y, t-i). \quad (9.2)$$

Silhouettes describe the moving object in the scene, but as such, they do not describe any motion. Therefore, difference of silhouettes is used as the representation when constructing the MHI. Furthermore, this representation allows to use online weighting on different subareas of the MHI as described later in this section. Silhouettes are used for MEI calculation to obtain a better overall description of the human pose. Figure 9.1 illustrates the templates.

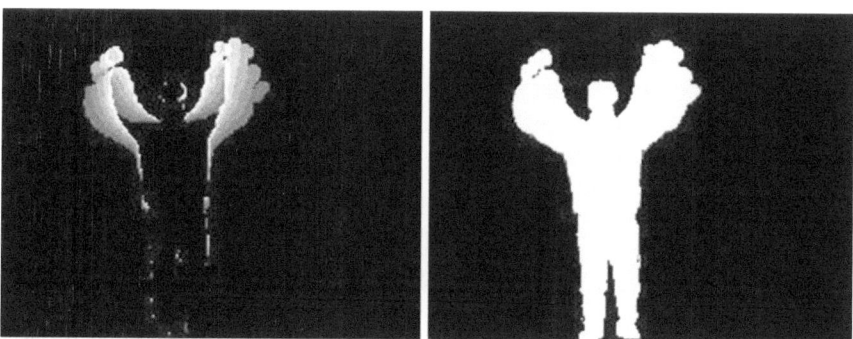

Fig. 9.1 Illustration of the MHI (*left*) and MEI (*right*) in a case where a person is raising both hands

When MHI and MEI are used to represent an action as a whole, setting the duration parameter τ is critical. This is not always easy as the duration of different actions as well as different instances of the same action can vary a lot. The problem with the temporal template representation is that actions that occupy the same space at different times cannot be modeled properly as the observed features will overlap and new observations will erase old ones. This problem is solved by fixing τ to give a short term motion representation and modeling the actions as a sequence of templates. The sequential modeling is then done with HMMs.

LBP histograms are used to characterize both MHI and MEI. This provides a new texture based descriptor of human movements that describes human movements on two levels. From the definition of the MHI and MEI it can be seen that the LBP codes from MHI encode the information about the direction of motion whereas the MEI based LBP codes describe the combination of overall pose and shape of motion.

It should be noted that some areas of MHI and MEI contain more information than others when texture is considered. MHI represents motion in gray level changes, which means that the outer edges of MHI may be misleading. In these areas there is no useful motion information and so the non-moving pixels having zero value should not be included in the calculation of the LBP codes. Therefore, calculation of LBP features is restricted to the non-monotonous area within the MHI template. For MEI the case is just the opposite, the only meaningful information is obtained around the boundaries.

Also, the LBP histogram of an image only contains information about the local spatial structures and does not give any information about the overall structure of motion. To preserve the rough structure of motion the MHI and MEI is divided into subregions. In the present approach the division into four regions is done through the centroid of the silhouette. This division roughly separates the limbs. For many actions seen from the side view, for example sitting down, the division does not have any clear interpretation but it preserves the essential information about the movements. This choice of division may not be optimal, and by choosing a more specified division scheme, one could increase the resolution of the description and model more specific activities.

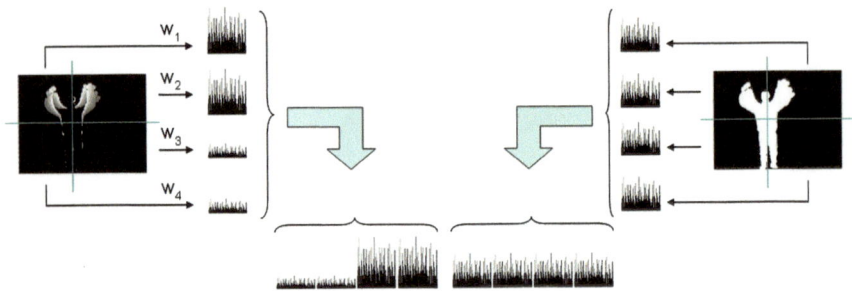

Fig. 9.2 Illustration of the formation of the feature histogram. In this frame the *top two subimages* in MHI have high weights compared to the *bottom two*

In many cases some of the MHI subregions contain much more motion than the others and thus provide more information. To give more focus on more meaningful areas of the images, spatial enhancement is performed by assigning different weights to the subregions. Instead of using prior weights, weights are given online based on the relative amount of motion the subimage contains. The weights are given as the ratio of the area of nonzero pixels that the MHI subregion contains to the area of nonzero pixels in the whole image. An example of how the weights are assigned is illustrated in Fig. 9.2. As MEI describes the pose of the person and all parts have similar importance, all subregion histograms for MEI are given equal weights.

Finally, to form a description of a frame, all the MHI and MEI based LBP subregion histograms are concatenated into one global histogram and normalized so that the sum of the histogram equals one. Figure 9.2 illustrates the MHI and MEI, their division into subregions and the formation of LBP histograms. The sequential development of the features is modeled with HMMs.

9.3 Dynamic Texture Method for Motion Description

In this section a dynamic texture based approach for human action description is presented. Instead of using a method like MHI to incorporate time into the description, the dynamic texture features capture the dynamics straight from image data. In the following subsections it is shown how the LBP-TOP features can be used to perform background subtraction to locate a bounding volume of human in xyt space and how the same features can be used for describing human movements.

9.3.1 Human Detection with Background Subtraction

Background subtraction is used as the first stage of processing in many approaches to human action recognition. Therefore, it can have a huge affect on the performance of such a system if the location and shape of the person need to be accurately determined. Also, from application point of view, the high computational cost and memory demands of many background subtraction methods may limit their use in systems requiring processing at video rate. The problem of computation cost is tackled by using the same features for both human detection and action recognition. Furthermore, this dynamic texture approach is designed so that an accurate silhouette is not needed and a bounding box is sufficient.

Usually background subtraction is done by modeling the pixel color and intensities [15, 28]. A different kind of approach was presented in Chap. 8, which introduced a region based method that uses LBP features from a local neighborhood. Unlike these, the image plane itself is not considered here but instead the temporal planes xt and yt. A purpose was to show the descriptive power of the dynamic texture features and therefore only focus on the temporal planes. This does not mean that temporal planes alone are better than xy for background subtraction; instead it was demonstrated that they are suitable for the task. The temporal planes are quite interesting from background subtraction point of view as they can encode both motion and some low level structural information.

In the presented method, a background model consists of a codebook C for every local neighborhood. When pixel values of a neighborhood are close to one another, the thresholding operation in the LBP feature extraction can be vulnerable to noise. In Chap. 8 [6], a bias was added to the center pixel, i.e., the term $s(g_p - g_c)$ in Eq. 2.10 is replaced with the term $s(g_p - g_c + a)$. This idea is also adopted here and neighborhood specific bias values are used. Thus, the background model consists of codebook C and the bias a for each pixel for both the temporal planes.

Overlapping volumes of duration $\Delta t = 2R_t + 1$ are used as an input to the method, i.e., for the current frame there is a buffer of Rt frames before and after the frame from which the spatiotemporal features are calculated. A pixel in the current frame is determined to belong to an object if the observed LBP code of a pixel neighborhood of the input volume does not match the codes in the corresponding codebook. The result from xt and yt planes can be combined using the logical AND operator. With this method, one can extract the bounding volume (3D equivalent to a bounding box in 2D) of a human in each space time volume and then use the volumes for action recognition as described in the next subsection. For the current application, the background model is not adaptive and the model must be learned by observing an empty scene. Preliminary experimental results are presented in Sect. 9.4. This method can be extended to be adaptive to changes in the background.

9.3.2 Action Description

The dynamic LBP-TOP features, previously introduced in Chap. 3, are used for human action description and the input for action recognition are the dynamic LBP-TOP features extracted from the detected human xyt volumes. Notice that when the above background subtraction is used, the LBP-TOP features have already been

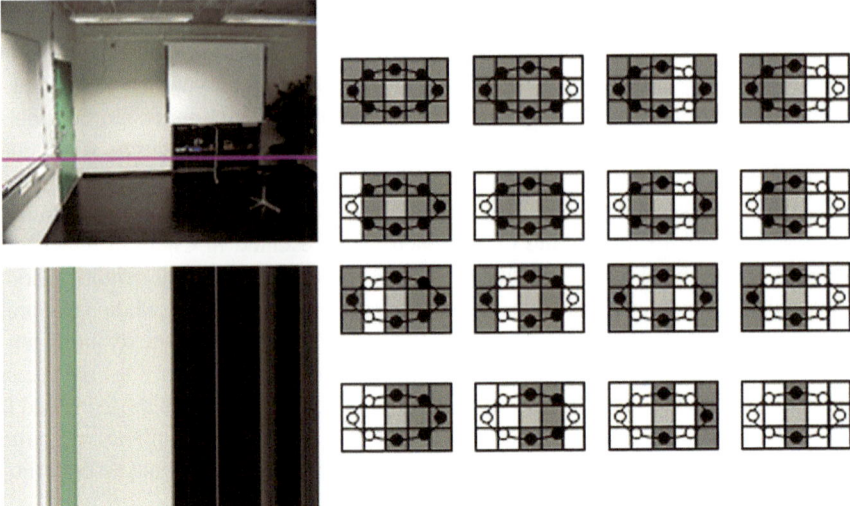

Fig. 9.3 Illustration of LBP patterns that represent no motion. The *two images on the left* illustrate a state a static scene containing no motion and a resulting *xt* plane. The *bit patterns* illustrate the resulting codes that do not describe any motion. Consider nearest neighbor interpolation for the simplicity of the illustration and also note the bias on the center pixel

computed and the action description can be obtained very efficiently. However, a couple of points need addressing to enhance the performance of the features.

As the silhouette data are not used but images from a bounding volume that contains the human, the input also contains some background information as well as the static human body parts. These regions do not contain any useful motion information and only describe the structure of the scene and the clothing of the person.

Considering the images illustrated in Fig. 9.3, static parts of the images produce stripe like patterns for the *xt* and *yt* planes. It can be observed that certain LBP codes represent these stripes. On *xt* and *yt* planes, symmetrical LBP kernels (kernels with even number of sampling points) have sampling point pairs that lie on the same spatial location but at different times (there are three such pairs in the eight-point kernel as can be seen in Fig. 9.3). It can be seen that the common property of these codes is that both bits of each pair are the same. If the bits of a pair are different for an observed code, in ideal case (no noise) this must be because of motion in the scene. As it is wished to obtain a motion description, the stripe pattern codes (SPC(P)) that contain no motion information are removed from the histogram. The stripe patterns are always the same for a given LBP kernel and only their relative appearance frequency depends on the scene structure. Cutting off these bins reduces the histogram length for an eight point neighborhood into 240 bins instead of 256, but more importantly, it also improves the motion description.

Similarly to the static LBP representation, the dynamic LBP features calculated over the whole bounding volume encode only the local properties of the movements

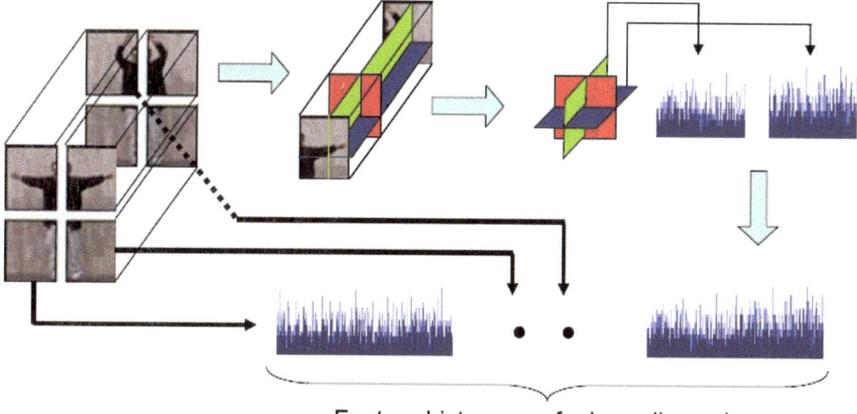

Feature histogram of a bounding volume

Fig. 9.4 Illustration of the formation of the feature histogram from a bounding volume

without any information about their local or temporal locations. For this reason the detected bounding volume is divided through its center point into four regions and form global feature histogram by concatenating the subvolume histograms. Using the subvolume representation the motion on three different levels is encoded: pixel-level (single bins in the subvolume histogram), region-level (whole subvolume histogram) and global-level (concatenated subvolume histograms).

The subvolume division and the formation of the feature histogram are illustrated in Fig. 9.4. All the subvolume histograms are concatenated and the resulting histogram is normalized by setting its sum equal to one. This is the representation of a bounding volume. The final histogram length in the representation is $N_s \times (2^{P_{xt}} - \text{SPC}(P_{xt}) + 2^{P_{yt}} - \text{SPC}(P_{yt}))$, where N_s is the number of subvolumes. For example, $P_{xt} = P_{yt} = 8$ with four subvolumes gives $4 \times (240 + 240) = 1920$ features. Finally the temporal development of the features is modeled using HMMs.

9.3.3 Modeling Temporal Information with Hidden Markov Models

As described before, short term human movements are described using histograms of local features extracted from a space time representation. The temporal development of the histograms is then modeled with HMM. The models are briefly described next but see tutorial [23] for more details on HMMs. In the presented approach a HMM that has N states $Q = q_1, q_2, \cdots, q_N$ is defined with the triplet $\lambda = (A, \pi, H)$, where A is the $N \times N$ state transition matrix, π is the initial state distribution vector and the H is the set of output histograms.

The probability of observing an LBP histogram h_{obs} is the texture similarity between the observation and the model histograms. Histogram intersection was chosen

Fig. 9.5 Illustration of the temporal development of the feature histograms and a simple three state circular HMM

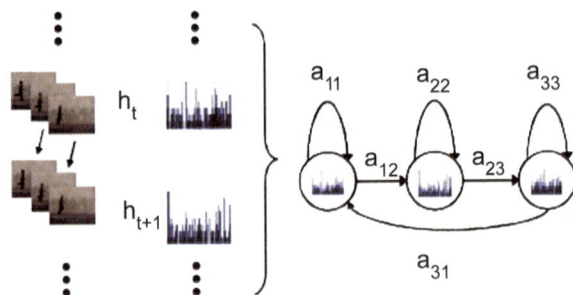

as the similarity measure as it satisfies the probabilistic constraints. Furthermore, a penalty factor n is used that allows to adjust the tightness of the models. Thus, the probability of observing h_{obs} in state i at time t is given as:

$$P(h_{obs}|s_t = q_i) = \left[\sum \min(h_{obs}, h_i)\right]^n, \tag{9.3}$$

where s_t is the state at time step t, and h_i is the observation histogram in state i. The summation is done over the bins. It can be seen that histograms with a larger deviation from the model are penalized more with a larger n.

Figure 9.5 illustrates how the features are calculated as time evolves and a simple circular HMM. Different kind of model topologies can be used: circular models are suitable for modeling repetitious movements like walking and running, whereas left-to-right models are suitable for movements like bending for example. HMMs can be used for activity classification by training an HMM for each action class. A new observed unknown feature sequence $H_{obs} = h_{obs1}, h_{obs2}, \cdots, h_{obsT}$ can be classified as belonging to the class of the model that maximizes $P(H_{obs}|\lambda)$, i.e., the probability of observing H_{obs} from the model λ.

9.4 Experiments

The performance of the method is demonstrated by experimenting with the Weizmann [25] and KTH datasets [24]. The datasets have become popular benchmark databases [22, 25, 29, 31], so the results can be directly compared to others reported in the literature.

The *Weizmann dataset* consists of ten different activities performed by nine different persons. Figure 9.6 illustrates the activities.

The action classification experiments were then performed on the Weizmann dataset using HMM modeling. The HMM topologies were set to circular for all models. A classification accuracy of 98.9% was obtained using both the static and the dynamic texture based method. The parameters are the same as used above. HMMs with seven states were used and $n = 7$, though both these values can be changed without too much of an affect on the performance.

Fig. 9.6 Illustration of the movement classes in the Weizmann database. Starting from the upper left corner the movements are: Bending, Jumping jack, Jumping, Jumping in place ('Pjump'), Gallop sideways ('Side'), Running, Skipping, Walking, Wave one hand ('wave1') and Wave two hands ('wave2')

It should be noted that the methods use different input data: the static texture based method uses silhouettes, whereas the dynamic texture method uses image data. Therefore, an experiment with dynamic texture based method but with silhouettes as input was also performed. Using the same parameters as above, 96.6% classification accuracy was achieved. This result is slightly lower than the result achieved with static texture method. It is interesting that the dynamic texture based method works better when images are used as input. This is probably because texture statistics extracted from image data is more reliable than the statistics taken from silhouettes where one can get interesting features only around the edges.

SVM classification with the histograms that were used earlier for feature analysis was also tried. The classification accuracies of 95.6% and 97.8% were achieved using dynamic (both on image and silhouette data) and static texture, respectively. Interestingly there is not much difference in the results between SVM and HMM modeling even though the input of the SVM does not contain the sequential information. This shows that the used histograms of local features capture much of the movement dynamics.

Results achieved by others on this database are summarized in Table 9.1. From the image based approaches Boiman and Irani [3] report the best overall recognition result, but their test set does not include the *Skipping* class. This extra class causes the only mistake made in the test set when using HMM and the image based dynamic texture features. This test was also run without the skipping class and in this case the presented methods were able to classify all of the movements correctly. The LBP-based methods give the best results on the database when image data is used as an input and are also very competitive against approaches that are based on silhouette data. As can be seen in Table 9.1, in general, silhouette based methods work better but new and effective image based methods are emerging. It can be predicted that the image based approaches have the potential to outperform the silhouette based methods in the future.

The *KTH dataset* consists of 25 people performing six actions in four different conditions. The actions in the database are *Boxing*, *Handwaving*, *Handclapping*, *Jogging*, *Running* and *Walking*. Each person performs these actions under four dif-

Table 9.1 Results reported in the literature for the Weizmann database. The columns represent the reference, input data type, number of activity classes, number of sequences and finally the classification result

Reference	Input	Act.	Seq.	Res.
Dynamic texture method (HMM)	**image data**	**10 (9)**	**90 (81)**	**98.9% (100%)**
Dynamic texture method (SVM)	**image data**	**10**	**90**	**95.6%**
Scovanner et al. 2007 [25]	image data	10	92	82.6%
Boiman and Irani 2006 [3]	image data	9	81	97.5%
Niebles et al. 2007 [21]	image data	9	83	72.8%
Static texture method (HMM)	**silhouettes**	**10 (9)**	**90 (81)**	**98.9% (98.7%)**
Static texture method (SVM)	**silhouettes**	**10**	**90**	**97.8%**
Dynamic texture method (HMM)	**silhouettes**	**10**	**90**	**96.6%**
Dynamic texture method (SVM)	**silhouettes**	**10**	**90**	**95.6%**
Wang and Suter 2007 [29]	silhouettes	10	90	97.8%
Ikizler and Duygulu 2007 [7]	silhouettes	9	81	100%

Fig. 9.7 Illustration of the six action classes in the KTH dataset and the different capturing conditions s1–s4

ferent conditions: outdoors, outdoors with scale variations, outdoors with clothing variation and indoors. Figure 9.7 illustrates the variance in the data.

The videos in the database are recorded with a hand held camera and contain zooming effects so using simple background subtraction methods does not work. Instead, the human bounding box was manually marked in a few frames for every movement and the bounding box was interpolated for the other frames. For *Jogging*, *Running* and *Walking* movements where the person is not always in the field of view, the bounding boxes were marked so that in the beginning and end of movements the person was still partially out of view. It should be noted that the extraction of

Table 9.2 Comparison of results achieved on the KTH dataset

Reference	Result
Kim et al. (2007) [14]	95.3%
Dynamic texture method	**93.8%**
Wong et al. (2007) [31]	91.6%
Static texture method (MHI)	90.8%
Niebles et al. (2008) [22]	81.5%
Ke et al. (2007) [9]	80.9%

bounding boxes could also be done automatically using a tracker or human detector. Since it is not possible to get the silhouettes frame differencing was used to build the MHI description for static texture based method. The use HMM modeling did not provide satisfactory results. This is due to the huge variation in the data which the used simple HMMs with the intersection similarity measure cannot capture. Instead the nearest neighbor and SVM classification were applied.

Experiments were performed using leave one out method for the database. The center point of the bounding box was used for the spatial division into subregions (subvolumes for the dynamic texture). Then the static texture ($R = 3$, $P = 8$, $\tau = 4$) and dynamic LBP-TOP ($R_t = 3$, $R_x = 3$, $R_y = 3$, $P_{xt} = P_{yt} = 8$ and constant bias $a = 5$) features were extracted from the whole duration of the action (varies from 13 to 362 frames), the histograms were concatenated and normalized over the duration of the action. A larger radius was chosen in both space and time to better capture the movement dynamics as the classification methods do not consider sequential information.

Using nearest neighbor classification accuracies of 85.9% and 89.8% were achieved for static and dynamic texture methods, respectively. Similarly, using an SVM with RBF kernel provided accuracies of 90.8% and 93.8%. Only the MHI based features were used for the static texture based method. In the absence of silhouettes, the MEI based features lowered the results considerably. It can be noticed that although both of the methods performed equally well on the easier Weizmann dataset, the more advanced dynamic texture based method outperforms the more naive static texture based method in this more challenging experiment.

The recent results reported in the literature [9, 14, 22, 31] for this database vary between 80%–95%. Table 9.2 summarizes the results achieved by others with a similar leave one out test setup. It can be seen that the result for the presented method is among the best.

9.5 Discussion

In this chapter the ideas of spatiotemporal analysis and the use of local features for human movement description were adopted. Two texture based methods were presented for human action recognition that naturally combine motion and appearance cues. The first uses temporal templates to capture movement dynamics and then uses texture features to characterize the observed movements. This idea was then extended into a spatiotemporal space by describing human movements with dynamic texture features. Following recent trends in computer vision, the latter method is designed to work with image data rather than silhouettes.

By using local properties of human motion, the LBP-based methods are robust against variations in performing actions as the high accuracy in the experiments on the Weizmann and KTH datasets shows. They are computationally simple and suitable for various applications.

It was also noticed in experiments that the dynamic texture based method works better when images are used as input instead of silhouettes. Furthermore, the image based dynamic texture method outperforms the simpler static texture based method on the more challenging KTH dataset. It can be concluded that image based approaches have the potential to outperform the silhouette based methods that have dominated the action recognition field in the past.

Recently, an approach for human gait recognition that inherently combines appearance and motion was developed [12]. The LBP-TOP descriptors presented in Chap. 3 are used to describe human gait in a spatiotemporal way. A new coding of multiresolution uniform local binary patterns was proposed and used in the construction of spatiotemporal LBP histograms. The suitability of this representation for gait recognition was tested on the popular CMU MoBo dataset, obtaining excellent results in comparison to the state of the art methods.

LBP-TOP has recently become popular in action recognition area. There are many developments on the basis of it. Ma and Cisar [18] presented a method using dynamic texture descriptor for event detection. Image sequences are divided into several regions, then a flow is formed on the basis of the similarity of the LBP-TOP features on the regions. Their results on real dataset are promising. On the basis of the work in this chapter, Yeffet and Wolf [32] proposed Local Trinary Patterns, which combine Local Binary Patterns with the appearance invariance and adaptability of patch matching based methods for human action recognition. Mattivi and Shao [19] applied LBP-TOP as a sparse descriptor to activity recognition as well. They first detect space-time interest points and then describe a video sequence using LBP-TOP as a collection of spatial-temporal words. Later they evaluated and compared different feature detection and feature description methods for part-based approaches in human action recognition [26], in which Extended Gradient LBP-TOP obtained promising results.

References

1. Blank, M., Gorelick, L., Shechtman, E., Irani, M., Basri, R.: Actions as space-time shapes. In: Proc. International Conference on Computer Vision, vol. 2, pp. 1395–1402 (2005)
2. Bobick, A., Davis, J.: The recognition of human movement using temporal templates. IEEE Trans. Pattern Anal. Mach. Intell. **23**(3), 257–267 (2001)
3. Boiman, O., Irani, M.: Similarity by composition. In: Proc. Neural Information Processing Systems Conference, p. 8 (2006)

4. Dollár, P., Rabaud, V., Cottrell, G., Belongie, S.: Behavior recognition via sparse spatio-temporal features. In: Proc. International Workshop on Visual Surveillance and Performance Evaluation of Tracking and Surveillance, pp. 65–72 (2005)
5. Gavrila, D.M.: The visual analysis of human movement: A survey. Comput. Vis. Image Underst. **73**(3), 82–98 (1999)
6. Heikkilä, M., Pietikäinen, M.: A texture-based method for modeling the background and detecting moving objects. IEEE Trans. Pattern Anal. Mach. Intell. **28**(4), 657–662 (2006)
7. Ikizler, N., Duygulu, P.: Human action recognition using distribution of oriented rectangular patches. In: Proc. ICCV Workshop on Human Motion Understanding, Modeling, Capture and Animation, pp. 271–284 (2007)
8. Ke, Y., Sukthankar, R., Hebert, M.: Efficient visual event detection using volumetric features. In: Proc. International Conference on Computer Vision, pp. 165–173 (2005)
9. Ke, Y., Sukthankar, R., Hebert, M.: Spatio-temporal shape and flow correlation for action recognition. In: Proc. IEEE Conference on Computer Vision and Pattern Recognition, pp. 1–8 (2007)
10. Kellokumpu, V., Zhao, G., Pietikäinen, M.: Human activity recognition using a dynamic texture based method. In: Proc. British Machine Vision Conference (2008)
11. Kellokumpu, V., Zhao, G., Pietikäinen, M.: Texture based description of movements for activity analysis. In: Proc. International Conference on Computer Vision Theory and Applications, vol. 1, pp. 206–213 (2008)
12. Kellokumpu, V., Zhao, G., Pietikäinen, M.: Dynamic texture based gait recognition. In: Advances in Biometrics. Lecture Notes in Computer Science, vol. 5558, pp. 1000–1009. Springer, Berlin (2009)
13. Kellokumpu, V., Zhao, G., Pietikäinen, M.: Recognition of human actions using texture descriptors. Machine Vision and Applications (2011). doi:10.1007/s00138-009-0233-8
14. Kim, K., Wong, S., Cipolla, R.: Tensor canonical correlation analysis for action classification. In: Proc. IEEE Conference on Computer Vision and Pattern Recognition, pp. 1–8 (2007)
15. Kim, K., Chalidabhongse, T.H., Harwood, D., Davis, L.S.: Background modeling and subtraction by codebook construction. In: Proc. IEEE International Conference on Image Processing, pp. 3061–3064 (2004)
16. Kobyashi, T., Otsu, N.: Action and simultaneous multiple-person identification using cubic higher-order auto-correlation. In: Proc. International Conference on Pattern Recognition, vol. 4, pp. 741–744 (2004)
17. Laptev, I., Lindeberg, T.: Space-time interest points. In: Proc. International Conference on Computer Vision, vol. 1, pp. 432–439 (2003)
18. Ma, Y., Cisar, P.: Event detection using local binary pattern based dynamic textures. In: Proc. CVPR Workshop on International Conference on Spoken Language Proceedings, pp. 38–44 (2009)
19. Mattivi, R., Shao, L.: Human action recognition using LBP-TOP as sparse spatio-temporal feature descriptor. In: Proc. International Conference on Computer Analysis of Images and Patterns, pp. 740–747 (2009)
20. Moeslund, T.B., Hilton, A., Krüger, V.: A survey of advances in vision-based human motion capture and analysis. Comput. Vis. Image Underst. **104**, 90–126 (2006)
21. Niebles, J.C., Fei-Fei, L.: A hierarchical model of shape and appearance for human action classification. In: Proc. IEEE Conference on Computer Vision and Pattern Recognition, pp. 1–8 (2007)
22. Niebles, J.C., Wang, H., Fei-Fei, L.: Unsupervised learning of human action categories using spatial-temporal words. Int. J. Comput. Vis. **79**(3), 299–318 (2008)
23. Rabiner, L.R.: A tutorial on hidden Markov models and selected applications in speech recognition. Proc. IEEE **77**(2), 257–285 (1989)
24. Schuldt, C., Laptev, I., Caputo, B.: Recognizing human actions: A local SVM approach. In: Proc. International Conference on Pattern Recognition, pp. 32–36 (2004)
25. Scovanner, P., Ali, S., Shah, M.: A 3-dimensional SIFT descriptor and its application to action recognition. In: Proc. ACM Multimedia, pp. 357–360 (2007)

26. Shao, L., Mattivi, R.: Feature detector and descriptor evaluation in human action recognition. In: Proceedings of the ACM International Conference on Image and Video Retrieval, pp. 477–484 (2010)
27. Shechtman, E., Irani, M.: Space-time behavior based correlation. In: Proc. IEEE Conference on Computer Vision and Pattern Recognition, pp. 405–412 (2005)
28. Stauffer, C., Grimson, W.E.L.: Adaptive background mixture models for real-time tracking. In: Proc. IEEE Conference Computer Vision and Pattern Recognition, vol. 2, pp. 246–252 (1999)
29. Wang, L., Suter, D.: Recognizing human activities from silhouettes: Motion subspace and factorial discriminative graphical model. In: Proc. IEEE Conference on Computer Vision and Pattern Recognition, pp. 1–8 (2007)
30. Weinland, D., Ronfard, R., Boyer, E.: Free viewpoint action recognition using motion history volumes. Comput. Vis. Image Underst. **104**(2–3), 249–257 (2006)
31. Wong, S., Kim, T., Cipolla, R.: Learning motion categories using both semantic and structural information. In: Proc. IEEE Conference on Computer Vision and Pattern Recognition, pp. 1–6 (2007)
32. Yeffet, L., Wolf, L.: Local trinary patterns for human action recognition. In: Proc. International Conference on Computer Vision, pp. 492–497 (2009)
33. Yilmaz, A., Shah, M.: Action sketch: A novel action representation. In: Proc. IEEE Conference on Computer Vision and Pattern Recognition, vol. 1, pp. 984–989 (2005)

Part IV
Face Analysis

Chapter 10
Face Analysis Using Still Images

Automatic face analysis has become a very active topic in computer vision research as it is useful in several applications, like biometric identification, visual surveillance, human-machine interaction, video conferencing and content-based image retrieval. Face analysis may include face detection and facial feature extraction, face tracking and pose estimation, face and facial expression recognition, and face modeling and animation [18, 54]. All these tasks are challenging due to the fact that facial appearance varies due to changes in pose, expression, illumination and other factors such as age and make-up. Therefore, one should derive facial representations that are robust to these factors.

Recent developments showed that the local binary patterns (LBP) provide outstanding results in representing and analyzing faces. This chapter explains how to easily derive efficient LBP based face descriptions which combine into a single feature vector the global shape and local texture of a facial image. The obtained representations are then applied to face and eye detection, face recognition, and facial expression recognition, yielding in excellent performance.

Face analysis is perhaps the most fascinating application of LBP. While texture features have been successfully used in various computer vision applications, only relatively few works have considered them in facial image analysis before the introduction of the LBP based face representations in 2004 [1, 4]. Since then, the methodology has inspired plenty of new methods in face analysis, thus revealing that texture based region descriptors can be very efficient in representing and analyzing facial patterns. This chapter gives also a review of some interesting works on face analysis using or inspired by LBP.

10.1 Face Description Using LBP

In the LBP approach for texture classification (Chap. 4), the occurrences of the LBP codes in an image are collected into a histogram. The classification is then performed by computing simple histogram similarities. However, considering a similar approach for facial image representation results in a loss of spatial information and

M. Pietikäinen et al., *Computer Vision Using Local Binary Patterns*,
Computational Imaging and Vision 40,
DOI 10.1007/978-0-85729-748-8_10, © Springer-Verlag London Limited 2011

Fig. 10.1 Example of
an LBP based facial
representation

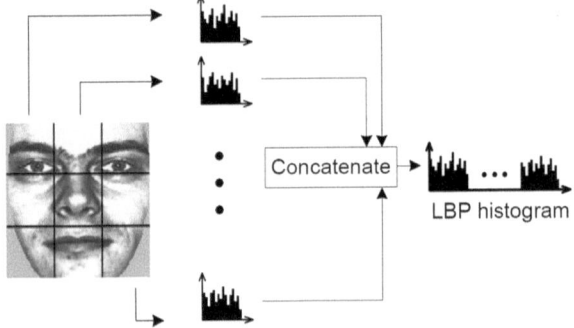

therefore one should codify the texture information while retaining also their locations. One way to achieve this goal is to use the LBP texture descriptors to build several local descriptions of the face and combine them into a global description. Such local descriptions have been gaining interest lately which is understandable given the limitations of the holistic representations. These local feature based methods seem to be more robust against variations in pose or illumination than holistic methods.

Another reason for selecting the local feature based approach is that trying to build a holistic description of a face using texture methods is not reasonable since texture descriptors tend to average over the image area. This is a desirable property for textures, because texture description should usually be invariant to translation or even rotation of the texture and, especially for small repetitive textures, the small-scale relationships determine the appearance of the texture and thus the large-scale relations do not contain useful information. For faces, however, the situation is different: retaining the information about spatial relations is important.

The basic methodology for LBP based face description is as follows: The facial image is divided into local regions and LBP texture descriptors are extracted from each region independently. The descriptors are then concatenated to form a global description of the face, as shown in Fig. 10.1.

The basic histogram that is used to gather information about LBP codes in an image can be extended into a *spatially enhanced histogram* which encodes both the appearance and the spatial relations of facial regions. As the facial regions $R_0, R_1, \ldots, R_{m-1}$ have been determined, the spatially enhanced histogram is defined as

$$H_{i,j} = \sum_{x,y} I\left\{f_l(x,y) = i\right\} I\left\{(x,y) \in R_j\right\}, \quad i = 0, \ldots, n-1, j = 0, \ldots, m-1.$$

This histogram effectively has a description of the face on three different levels of locality: the LBP labels for the histogram contain information about the patterns on a pixel-level, the labels are summed over a small region to produce information on a regional level and the regional histograms are concatenated to build a global description of the face.

It should be noted that when using the histogram based methods the regions $R_0, R_1, \ldots, R_{m-1}$ do not need to be rectangular. Neither do they need to be of the same size or shape, and they do not necessarily have to cover the whole image. It is also possible to have partially overlapping regions. AdaBoost is commonly used for selecting optimal LBP settings such as the size, shape and location of the local regions.

This outlines the original LBP based facial representation [1, 4] that has already attained an established position in face analysis research as it has been adopted by many leading scientists and research groups around the world, and applied to various facial image analysis tasks. Some applications of the LBP-based facial representation are discussed below.

10.2   Eye Detection

Local binary patterns have been used for eye detection is several works. For instance, inspired by the works of Viola and Jones on the use of Haar-like features with integral images [44] and that of Heusch et al. on the use of LBP as a preprocessing step for handling illumination changes [12], a robust approach for eye detection using Haar-like features extracted from LBP images was proposed in [10]. The eye images were first filtered by LBP operator ($LBP_{8,1}$) and then Haar-like features were extracted and used with AdaBoost for building a cascade of classifiers.

During training, the "bootstrap" strategy was used to collect the negative examples. First, non-eye samples were randomly extracted from a set of natural images which do not contain eyes. Additionally, negative training samples were also extracted from facial regions because it has been shown that this enhances the performance of the system. An eye detector was then built and run to collect all those non-eye patterns that were wrongly classified as eyes. These samples were added to the non-eye set for re-training.

In total, the system was trained using 3116 eye patterns (positive samples) and 2461 non-eye patterns (negative samples). The system was evaluated on a database containing over 30000 frontal face images. The results were compared against those obtained by using Haar-like features and LBP features separately. Detection rates of 86.7%, 81.3% and 80.8% were obtained when considering LBP/Haar-like features, LBP only and Haar-like features only, respectively. Some detection examples, using the combined approach, are shown in Fig. 10.2 (some face images are from Cohn-Kanade facial expression database [17]). The results assessed the efficiency of combining LBP and Haar-like features (86.7%) while LBP and Haar-like features alone gave a lower performance.

In [38], Sun and Ma used LBP features as inputs to SVMs for eye detection. The authors trained two SVM classifiers: the first one is used for detecting open eyes while the second is trained for handling closed eyes. Experimental results using the CAS-PEAL database proved the feasibility of the proposed approach in detecting not only open eyes but also closed ones.

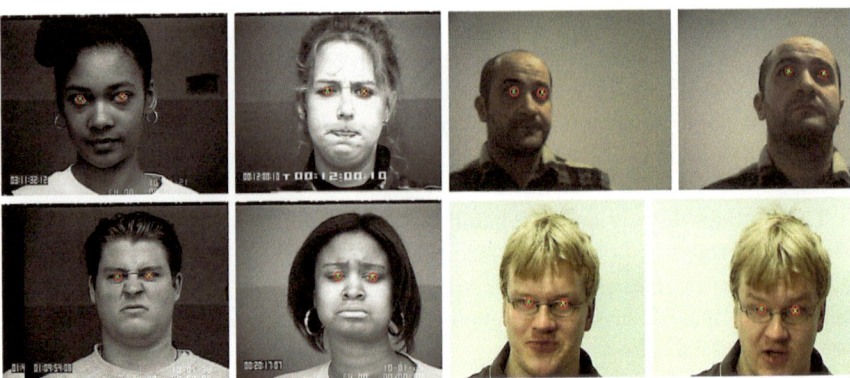

Fig. 10.2 Examples of eye detection results

In [22], an approach for eye gaze tracking using LBP features is proposed. LBP features are extracted from the eye images to encode the pupil-glint vector and a binocular vision approach is used to compute the space coordinates of the eye. The combined features are then fed into Support Vector Regression (SVR) for learning the gaze mapping function. The experimental results showed accurate gaze direction estimation.

10.3 Face Detection

Among the many proposed methods for detecting faces [48], appearance-based approaches which rely on training sets to capture the large variability in facial appearances have attracted much attention and demonstrated the best results. Generally, these methods scan an input image at all possible locations and scales, and then classify the sub-windows either as face or non-face. In many works, raw pixel data are used as features and fed to the classifiers. Such inputs can achieve surprisingly high recognition rates, but they are sensitive to all kinds of changes and require complex classifiers. Instead, Viola and Jones [44] used Haar-like features and the AdaBoost learning algorithm to overcome problems caused by the large number of extracted features. Due to their computational simplicity and discriminative power, the LBP features have recently been adopted for detecting faces yielding in excellent results outperforming many existing works.

For instance, Hadid et al. proposed an LBP based approach for face detection with very good results [9]. The authors noticed that the LBP based facial description presented in Sect. 10.1 and used for recognition in Sect. 10.4 is more adequate for larger-sized images. For example, in the FERET tests the images have a resolution of 130×150 pixels and were typically divided into 49 blocks, leading to a relatively long feature vector typically containing thousands of elements. However, in many applications such as face detection, the faces can be on the order of 20×20 pixels. Therefore, such representation cannot be used for detecting (or

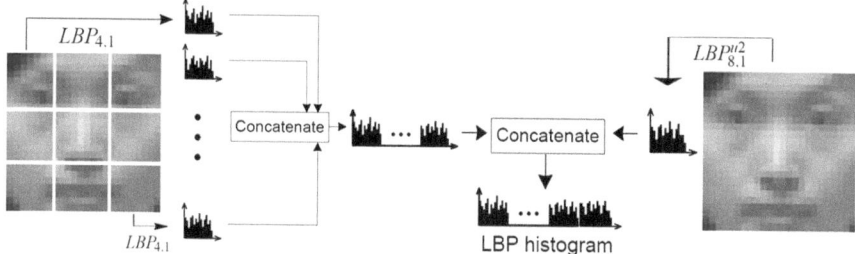

Fig. 10.3 Facial representation for low-resolution images: a face image is represented by a concatenation of a global and a set of local LBP histograms

even recognizing) low-resolution face images. The authors derived then a new LBP based representation which is suitable for low-resolution images and has a short feature vector needed for fast processing. A specific aspect of this representation is the use of overlapping regions and a 4-neighborhood LBP operator ($LBP_{4,1}$) to avoid statistical unreliability due to long histograms computed over small regions. Additionally, the holistic description of a face was enhanced by including the global LBP histogram computed over the whole face image.

Considering 19×19 as the basic resolution, the new LBP facial representation was derived as follows (see Fig. 10.3): A 19×19 face image is first divided into 9 overlapping regions of 10×10 pixels (overlapping size = 4 pixels). From each region, a 16-bin histogram using the $LBP_{4,1}$ operator is computed. The results are concatenated into a single 144-bin histogram. Additionally, $LBP_{8,1}^{u2}$ operator is applied to the whole 19×19 face image obtaining a 59-bin histogram which was added to the 144 bins previously computed. Thus, a $(59 + 144 = 203)$-bin histogram is obtained and used as the descriptor for face detection.

To assess the performance of the new representation, a face detection system using LBP features and an SVM was built. Given training samples (face and non-face images) represented by their extracted LBP features, an SVM classifier finds the separating hyperplane that has maximum distance to the closest points of the training set. These closest points are called *support vectors*. To perform a nonlinear separation, the input space is mapped into a higher dimensional space using Kernel functions. In the proposed approach, to detect faces in a given target image, a 19×19 subwindow scans the image at different scales and locations. A downsampling rate of 1.2 and a moving scan of 2 pixels were considered. At each iteration, the representation LBP(w) is computed from the subwindow and fed to the SVM classifier to determine whether it is a face or not (LBP(w) denotes the LBP feature vector representing the region scanned by the subwindow).

Additionally, given the results of the SVM classifier, a set of heuristics is performed to merge multiple detections and remove the false ones. For a given detected window, the detections within a neighborhood of 19×19 pixels (each detected window is represented by its center) are counted. The detections are removed if their number is less than 3. Otherwise, they are merged to keep only the one with the highest SVM output.

Fig. 10.4 Detection examples in several images from different sources. The images **c**, **d** and **e** are from the World Wide Web. Note: excellent detections of upright faces in **a**; detections under slight in-plane rotation in **a** and **c**; missed faces in **c**, **e** and **a** because of large in-plane rotation; missed face in **a** because of a pose-angled face; and a false detection in **e**

From the collected training sets, the proposed facial representations were first extracted. Then, these features were fed to the SVM classifier and were used to train the face detector. The system was run on several images from different sources to detect faces. Figures 10.4 and 10.5 show some detection examples. It can be seen that most of the upright frontal faces are detected. For instance, Fig. 10.5g shows perfect detections. In Fig. 10.5f, only one face is missed by the system. This miss is due to occlusion. A similar situation is shown in Fig. 10.4a in which the missed face is due to a large in-plane rotation. Since the system is trained to detect only in-plane rotated faces up to $\pm 18°$, it succeeded to find the slightly rotated faces in Fig. 10.4c, Fig. 10.4d and Fig. 10.5h and failed to detect largely rotated ones (as those in 10.4e and 10.4c). A false positive is shown in Fig. 10.4e while a false negative is shown in Fig. 10.4d. Notice that this false negative is expected since the face is pose-angled (i.e. not in frontal position). These examples summarize the main aspects of the LBP based detector on images from different sources.

In order to further investigate the performance gain using LBP features, another face detector was implemented using the same training and test sets. A similar SVM based face detector was considered but using different features as inputs. The normalized pixel features was chosen as inputs since it has been shown that such fea-

Fig. 10.5 Detection examples in several images from the subset of MIT-CMU tests. Note: excellent detections of upright faces in **f** and **g**; detection under slight in-plane rotation in **h**; missed face in **f** because of occlusion

tures perform better than the gradient and wavelet based ones when using with an SVM classifier [11]. The system was trained using the same training samples. The experimental results clearly showed the efficiency the LBP-based approach which compared favorably against the state-of-the-art algorithms. The results showed that: (i) the proposed LBP features are more discriminative than the normalized pixel values; (ii) the proposed representation is more compact as, for 19×19 face images, a 203-element feature vector was derived while the raw pixel features yield a vector of 361 elements; and (iii) the LBP based approach did not require histogram equalization and used a smaller number of support vectors. More details on these experiments can be found in [9].

The LBP based face detector in [9] was later extended with an aim to develop a real-time multi-view detector suitable for real world environments such as video surveillance, mobile devices and content based video retrieval [10]. The new approach uses LBP features in a coarse-to-fine detection strategy (pyramid architec-

Fig. 10.6 Examples of face detections on the CMU Rotated and Profile Test Sets

ture) embedded in a fast classification scheme based on AdaBoost learning. A real-time operation was achieved with as a good detection accuracy as the original, which was a much slower approach. The system handles out-of-plane face rotations in the range of $[-60°, +60°]$ and in-plane rotations in the range of $[-45°, +45°]$. As done by S. Li and Zhang in [19], the in-plane rotation is achieved by rotating the original images by $±30°$. Some detection examples on CMU Rotated and Profile Test Sets are shown in Fig. 10.6. The results are comparable to the state-of-the-art, especially in terms of detection accuracy. In terms of speed, the approach is slightly slower than the system proposed by S. Li and Zhang in [19].

In [15], a variant of LBP based facial representation, called "Improved LBP", was also adopted for face detection. In ILBP, the 3×3 neighbors of each pixel are not compared to the center pixel as in the original LBP, but to the mean value of the pixels. The authors argued that ILBP captures more information than LBP does. However, using ILBP, the length of the histogram increases rapidly. For instance, while $LBP_{8,1}$ uses a 256-bin histogram, $ILBP_{8,1}$ computes 511 bins. Using the ILBP features, the authors have considered a Bayesian framework for classifying the ILBP features. The face and non-face classes were modeled using multivariable Gaussian distributions while the Bayesian decision rule was used to decide on the "faceness" of a given pattern. The reported results are very encouraging. Later, the authors proposed another approach to face detection based on boosting ILBP features [16].

In [51], Zhang et al. also adopted boosting multi-block local binary pattern (MB-LBP) features for face detection. Their experiments showed that weak classifiers based on MB-LBP are more discriminative than Haar-like features and original LBP features. Combining ideas from Haar and LBP features have also given excellent results in accurate and illumination invariant face detection [31, 47].

Combining ideas from Haar and LBP features have also given excellent results in accurate and illumination invariant face detection [31, 47]. For instance, Roy and Marcel [31] proposed a new type of binary features called Haar Local Binary Pattern (HLBP) for fast and reliable face detection, especially in adverse imaging conditions. For computing the features, the bin values of LBP histograms calculated over two adjacent image sub-regions are compared. These sub-regions are similar to those in the Haar masks, hence the name of the features. They capture the region-specific variations of local texture patterns and are boosted using AdaBoost in a framework similar to that proposed by Viola and Jones. Experiments on several standard databases showed that HLBP features (combining the concepts of Haar feature with LBP) are able to model the region-specific variations of local texture and are relatively robust to wide variations in illumination, pose and background, and also slight variations in pose. The obtained results were significantly better than those obtained using Haar features only in adverse imaging conditions.

In [47], Yan et al. proposed another type of features combining Haar features and LBP for accurate face detection. The new features, called Locally Assembled Binary (LAB) Haar features, are inspired by Haar features by keeping only the ordinal relationship (named by binary Haar feature) rather than the difference between the accumulated intensities. Several neighboring binary Haar features are then assembled to capture their co-occurrence with similar idea to LBP. Experimental results on the CMU+MIT frontal face test set and CMU profile test set showed that the LAB features are more efficient than Haar and LBP features when used separately both in terms of discriminating power and computational cost.

10.4 Face Recognition

The original LBP based face representation introduced in 2004 by Ahonen et al. [1] was first assessed on the problem of face recognition. Since then, the methodology has inspired plenty of new methods. This section describes the application of LBP in the face recognition task. Typically a nearest neighbor classifier is used with LBP because the number of training (gallery) images per subject is generally low, often only one. The idea of a spatially enhanced histogram presented in Sect. 10.1 can be exploited further when defining the distance measure for the classifier. An indigenous property of the proposed face description method is that each element in the enhanced histogram corresponds to a certain small area of the face. Based on the psychophysical findings, which indicate that some facial features (such as eyes) play a more important role in human face recognition than other features [54], it

Fig. 10.7 (**a**) An example of a facial image divided into 7×7 windows. (**b**) The weights set for weighted χ^2 dissimilarity measure. *Black squares* indicate weight 0.0, *dark grey* 1.0, *light grey* 2.0 and *white* 4.0

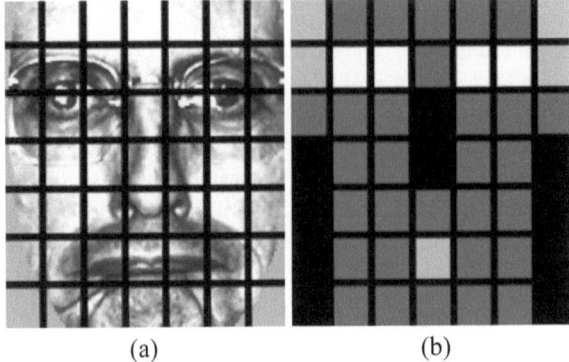

(a)   (b)

can be expected that in this method some of the facial regions contribute more than others in terms of extra-personal variance. Utilizing this assumption the regions can be weighted based on the importance of the information they contain. Figure 10.7 shows an example of weighting different facial regions [1]. Thus, the weighted chi square distance can be defined as

$$\chi_w^2(\mathbf{x}, \xi) = \sum_{j,i} w_j \frac{(x_{i,j} - \xi_{i,j})^2}{x_{i,j} + \xi_{i,j}}, \tag{10.1}$$

in which $\mathbf{x}$ and ξ are the normalized enhanced histograms to be compared, indices i and j refer to i-th bin in histogram corresponding to the j-th local region and w_j is the weight for region j.

The LBP based face recognition approach was extensively evaluated using the FERET face images [29]. The details of the experiments can be found in [1, 2, 4]. The recognition results (rank curves) are plotted in Fig. 10.8 [1]. The results clearly show that LBP approach yields higher recognition rates than the control algorithms (PCA [42], Bayesian Intra/Extrapersonal Classifier (BIC) [27] and Elastic Bunch Graph Matching EBGM [46]) in all the FERET test sets including changes in facial expression (*fb* set), lighting conditions (*fc* set) and aging (*dup I* & *dup II* sets). The results on the *fc* and *dup II* sets show that especially with weighting, the LBP based description is robust to challenges caused by lighting changes or aging of the subjects

To gain better understanding on whether the obtained recognition results are due to general idea of computing texture features from local facial regions or due to the discriminatory power of the local binary pattern operator, LBP approach was compared against three other texture descriptors, namely the gray-level difference histogram, homogeneous texture descriptor [25] and an improved version of the texton histogram [43]. The details of these experiments can be found in [2]. The results confirmed the validity of the LBP approach and showed that the performance of LBP in face description exceeds that of other texture operators LBP was compared to, as shown in Table 10.1. The main explanation for the better performance of the local binary pattern operator over other texture descriptors is its tolerance to monotonic gray-scale changes. Additional advantages are the computational efficiency of

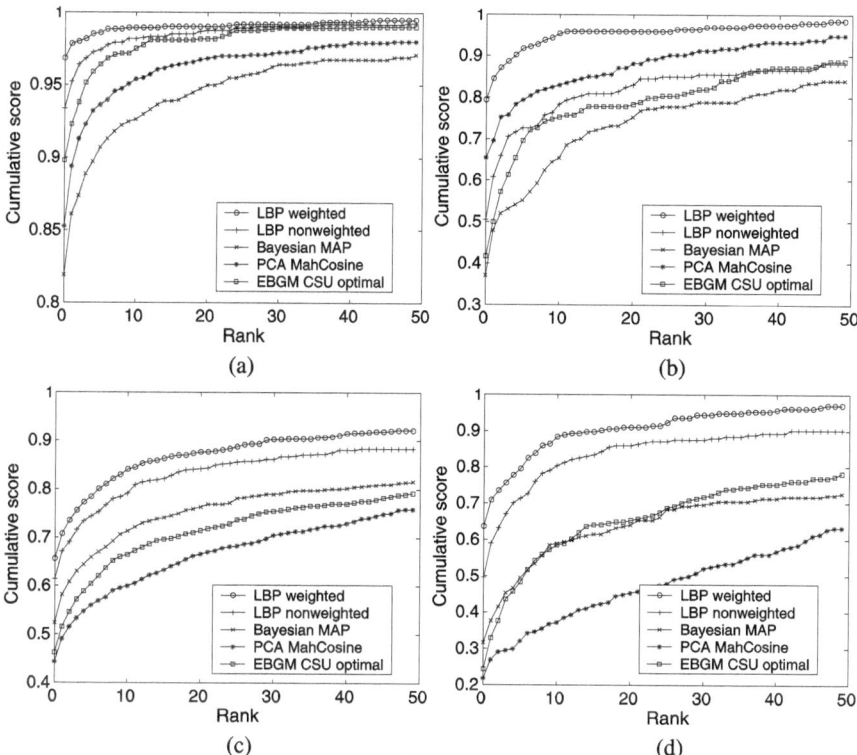

Fig. 10.8 The cumulative scores of the LBP and control algorithms on the (**a**) *fb*, (**b**) *fc*, (**c**) *dup I* and (**d**) *dup II* probe sets

Table 10.1 The recognition rates obtained using different texture descriptors for local facial regions. The first four columns show the recognition rates for the FERET test sets and the last three columns contain the mean recognition rate of the permutation test with a 95% confidence interval

Method	*fb*	*fc*	*dup I*	*dup II*	lower	mean	upper
Difference histogram	0.87	0.12	0.39	0.25	0.58	0.63	0.68
Homogeneous texture	0.86	0.04	0.37	0.21	0.58	0.62	0.68
Texton histogram	0.97	0.28	0.59	0.42	0.71	0.76	0.80
LBP (nonweighted)	0.93	0.51	0.61	0.50	0.71	0.76	0.81

the LBP operator and that no gray-scale normalization is needed prior to applying the LBP operator to the face image.

Recently, Tan and Triggs [40] developed a very effective preprocessing chain for face images and obtained excellent results using LBP-based face recognition for the FRGC database. Since then, many others have adopted this preprocessing chain for applications dealing with severe illumination variations. Figure 10.9 shows an example of gallery and probe images from the FRGC database and the corresponding

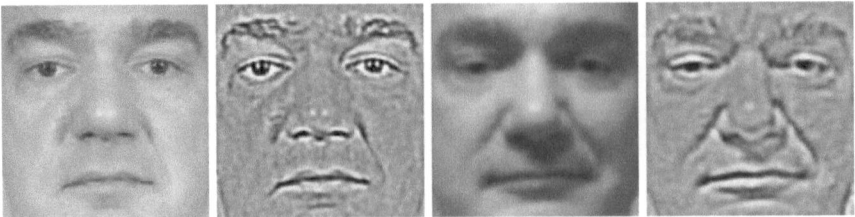

Fig. 10.9 Example of Gallery and probe images from the FRGC database, and their corresponding filtered images with Tan and Triggs' preprocessing chain

filtered images with Tan and Triggs' preprocessing method [40]. The authors also proposed a new facial representation based on Local Ternary Patterns for solving the problem caused by LBP's sensitivity to random and quantization noise. A distance transform-based similarity metric was used for decision. The method showed promising performance on three datasets with illumination variations.

The LBP based face representation has inspired plenty of other methods in face recognition. For instance, Chan et al. [6] considered multi-scale LBPs and derived new face descriptor from Linear Discriminant Analysis (LDA) of multi-scale local binary pattern histograms. The face image is first partitioned into several non-overlapping regions. In each region, multi-scale local binary uniform pattern histograms are extracted and concatenated into a regional feature. The features are then projected on the LDA space to be used as a discriminative facial descriptor. The method is tested in face identification on the standard FERET database and in face verification on the XM2VTS database with very promising results. In another work [5], the authors have also investigated color face recognition by extracting multispectral histograms using opponent color LBP operator and then projecting them into LDA space for recognition. Experiments in face verification on the XM2VTS and FRGC 2.0 databases showed results outperforming those of the state-of-the-art methods.

Zhang et al. [50] considered the LBP methodology for face recognition and used AdaBoost learning algorithm for selecting an optimal set of local regions and their weights. This yielded in a smaller feature vector length representing the facial images than that used in the original LBP approach [1]. However, no significant performance enhancement has been obtained. Later, Huang et al. [14] proposed a variant of AdaBoost called JSBoost for selecting the optimal set of LBP features for face recognition.

In order to deal with strong illumination variations, Li et al. developed a very successful system combining near-infrared (NIR) imaging with local binary pattern features and AdaBoost learning [20]. The invariance of LBP with respect to monotonic gray level changes makes the NIR images illumination invariant. The method achieved a verification rate of 90% at FAR = 0.001 and 95% at FAR = 0.01 on a database with 870 subjects.

The methods based on Gabor filtering and LBP provide complementary information for face analysis: LBP captures small and fine details, while Gabor filters encode appearance information over a broader range of scales. This has been successfully utilized in some recent studies, for example by Tan and Triggs [41] who extensively studied the benefits of combining Gabor wavelets and LBP features for face recognition. The proposed framework includes: robust photometric image normalization; separate feature extraction, PCA-based dimensionality reduction and scalar variance normalization of each modality; feature concatenation; Kernel DCA based extraction of discriminant nonlinear features; and finally cosine-distance based nearest neighbor classification in the KDCA reduced subspace. Experiments using several challenging face datasets including FRGC 1.0.4, FRGC 2.0.4 and FERET yielded very promising results and clearly indicated the complementary role of Gabor wavelets and LBP features.

Zhang et al. [52] proposed the extraction of LBP features from images obtained by filtering a facial image with 40 Gabor filters of different scales and orientations. Excellent results have been obtained on all the FERET sets. A downside of the method lies in the high dimensionality of the feature vector (LBP histogram) which is calculated from 40 Gabor images derived from each single original image. To overcome this problem of long feature vector length, Shan et al. [36] presented a new extension using Fisher Discriminant Analysis (FDA) instead of the χ^2 (chi-square) and histogram intersection which have been previously used in [52]. The authors constructed an ensemble of piecewise FDA classifiers, each of which is built based one segment of the high-dimensional LBP histograms. Impressive results were reported on the FERET database. In [53], Zhang et al. also demonstrated that Gabor phases are also useful for face recognition. The authors encoded Gabor phases through LBP and formed local feature histograms, yielding very good results on FERET database.

In [30], Rodriguez and Marcel proposed an approach based on adapted, client-specific LBP histograms for the face verification task. The method considers local histograms as probability distributions and computes a log-likelihood ratio instead of χ^2 similarity. A generic face model is considered as a collection of LBP histograms. Then, a client-specific model is obtained by an adaptation technique from the generic model under a probabilistic framework. The reported experimental results show that the proposed method yields good performance on two benchmark databases (XM2VTS and BANCA). Later, Ahonen and Pietikäinen [3] have further enhanced the face verification performance on the BANCA database by developing a novel method for estimating the local distributions of LBP labels. The method is based on kernel density estimation in xy-space, and it provides much more spatial accuracy than the earlier, block-based method of Rodriguez and Marcel [30].

LBP features have also been used in 3D face recognition. For instance. Li et al. [21] developed a face recognition system by fusing, both at feature and decision levels, 2D and 3D information extracted using LBP operator. AdaBoost was used for feature selection. Experiments on a database containing 252 subjects showed interesting results. In [28], Nanni and Lumini, have also adopted LBP to extract features from both 2D and 3D face images, yielding EER of 3.5% in experiments using a subset of 198 persons from the Notre-Dame database collection.

Several other LBP variants have been proposed and successfully applied to face recognition. Basically, many of the variants described in Chap. 2 have also been adopted to face recognition.

| Sadness | Disgust | Surprise | Happiness | Anger | Fear |

Fig. 10.10 Examples of images from the Japanese Female Facial Expression (JAFFE) database

10.5 Facial Expression Recognition

LBP has been also considered for facial expression recognition. For instance, in [7], an approach to facial expression recognition from static images was developed using LBP histograms computed over non-overlapping blocks for face description. The Linear Programming (LP) technique was adopted to classify seven facial expressions: anger, disgust, fear, happiness, sadness, surprise and neutral. During the training, the seven expression classes were decomposed into 21 expression pairs such as anger-fear, happiness-sadness etc. Thus, twenty-one classifiers were produced by the LP technique, each corresponding to one of the 21 expression pairs. A simple binary tree tournament scheme with pairwise comparisons was used for classifying unknown expressions. Good results (93.8%) were obtained for the Japanese Female Facial Expression (JAFFE) database used in the experiments. The database contains 213 images in which ten persons are expressing three or four times the seven basic expressions. Examples of images from the JAFFE database are shown in Fig. 10.10 [23]. Another approach to facial expression recognition using LBP features was proposed in [33]. Instead of the LP approach, template matching with weighted chi square statistic and SVM are adopted to classify the facial expressions using LBP features. Extensive experiments on the Cohn-Kanade database confirmed that LBP features are discriminative and more efficient than Gabor-based methods especially at low image resolutions. Boosting LBP features has also been considered for facial expression recognition in [32]. A comprehensive study on using LBP for facial expression recognition can be found in [35]. Gritti et al. studied the effect of face registration error on different approaches based on local features [8]. Histograms of Gradients (HOG), LBP, Local Ternary Patterns (LTP), and Gabor features were used as local features. They also considered the use of overlapping windows for LBP and LTP. A conclusion of this study was that LBP with overlapping gives the best recognition rate of 92.9% on the Cohn-Kanade database, while maintaining a compact feature vector. LBP with overlapping is also most robust against errors in the face registration.

Shan and Gritti [34], used AdaBoost for learning discriminative LBP histogram bins for facial expression recognition. Using multi-scale LBP bins, the approach yields recognition rate of 93.1% on the Cohn-Kanade database.

10.6 LBP in Other Face Related Tasks

The LBP approach has been also adopted to several other facial image analysis tasks such as gender recognition [37], demographic classification of age and ethnicity [49], iris recognition [39], head pose estimation [24] and 3D face recognition [21]. For instance, LBP is used in [13] with Active Shape Model (ASM) for localizing and representing facial key points since an accurate localization of such points of the face is crucial to many face analysis and synthesis problems such as face alignment. The local appearance of the key points in the facial images are modeled with an Extended version of Local Binary Patterns (ELBP). ELBP was proposed in order to encode not only the first derivation information of facial images but also the velocity of local variations. The experimental analysis showed that the combination ASM-ELBP enhances the face alignment accuracy compared to the original method used in ASM. Later, Marcel et al. [26] further extended the approach to locate facial features in images of frontal faces taken under different lighting conditions. Experiments on the standard and darkened image sets of the XM2VTS database assessed that the LBP-ASM approach gives superior performance compared to the basic ASM. Other works have also successfully combined LBP features with Gabor filters e.g. for age classification [45].

10.7 Conclusion

Face images can be seen as a composition of micro-patterns which can be well described by LBP texture operator. This observation was exploited to describe efficient face representations which have been successfully applied to various face analysis tasks, including face and eye detection, face recognition, and facial expression analysis problems. The extensive experiments have clearly shown the validity of LBP based face descriptions and demonstrated that texture based region descriptors can be very useful in nontraditional texture analysis tasks.

The success of LBP in face description is due to the discriminative power and computational simplicity of the LBP operator, and the robustness of LBP to monotonic gray scale changes caused by, for example, illumination variations. The use of histograms as features also makes the LBP approach robust to face misalignment and pose variations. For these reasons, the LBP methodology has already attained an established position in face analysis research. This is attested by the increasing number of works which adopted a similar approach.

References

1. Ahonen, T., Hadid, A., Pietikäinen, M.: Face recognition with local binary patterns. In: European Conference on Computer Vision. Lecture Notes in Computer Science, vol. 3021, pp. 469–481. Springer, Berlin (2004)

2. Ahonen, T., Pietikäinen, M., Hadid, A., Mäenpää, T.: Face recognition based on the appearance of local regions. In: Proc. International Conference on Pattern Recognition, vol. 3, pp. 153–156 (2004)
3. Ahonen, T., Pietikäinen, M.: Pixelwise local binary pattern models of faces using kernel density estimation. In: Proc. IAPR/IEEE International Conference on Biometrics, pp. 52–61 (2009)
4. Ahonen, T., Hadid, A., Pietikäinen, M.: Face description with local binary patterns: Application to face recognition. IEEE Trans. Pattern Anal. Mach. Intell. **28**(12), 2037–2041 (2006)
5. Chan, C.H., Kittler, J.V., Messer, K.: Multispectral local binary pattern histogram for component-based color face verification. In: Proc. IEEE Conference on Biometrics: Theory, Applications and Systems, pp. 1–7 (2007)
6. Chan, C.-H., Kittler, J., Messer, K.: Multi-scale local binary pattern histograms for face recognition. In: Proc. International Conference on Biometrics, pp. 809–818 (2007)
7. Feng, X., Pietikäinen, M., Hadid, A.: Facial expression recognition with local binary patterns and linear programming. Pattern Recognit. Image Anal. **15**(2), 546–548 (2005)
8. Gritti, T., Shan, C., Jeanne, V., Braspenning, R.: Local features based facial expression recognition with face registration errors. In: Proc. IEEE International Conference on Automatic Face and Gesture Recognition, pp. 1–8 (2008)
9. Hadid, A., Pietikäinen, M., Ahonen, T.: A discriminative feature space for detecting and recognizing faces. In: Proc. IEEE Conference on Computer Vision and Pattern Recognition, vol. 2, pp. 797–804 (2004)
10. Hadid, A., Zhao, G., Ahonen, T., Pietikäinen, M.: Face Analysis Using Local Binary Patterns. In: Mirmehdi, M., Xie, X., Suri, J. (eds.) Handbook of Texture Analysis, pp. 347–373. Imperial College Press, London (2008)
11. Heisele, B., Poggio, T., Pontil, M.: Face detection in still gray images. Technical Report 1687, Center for Biological and Computational Learning, MIT (2000)
12. Heusch, G., Rodriguez, Y., Marcel, S.: Local binary patterns as an image preprocessing for face authentication. In: Proc. International Conference on Automatic Face and Gesture Recognition, pp. 9–14 (2006)
13. Huang, X., Li, S.Z., Wang, Y.: Shape localization based on statistical method using extended local binary pattern. In: Proc. International Conference on Image and Graphics, pp. 184–187 (2004)
14. Huang, X., Li, S.Z., Wang, Y.: Jensen-Shannon boosting learning for object recognition. In: Proc. IEEE Conference on Computer Vision and Pattern Recognition, vol. 2, pp. 144–149 (2005)
15. Jin, H., Liu, Q., Lu, H., Tong, X.: Face detection using improved LBP under Bayesian framework. In: Proc. International Conference on Image and Graphics, pp. 306–309 (2004)
16. Jin, H., Liu, Q., Tang, X., Lu, H.: Learning local descriptors for face detection. In: Proc. IEEE International Conference on Multimedia and Expo, pp. 928–931 (2005)
17. Kanade, T., Cohn, J.F., Tian, Y.: Comprehensive database for facial expression analysis. In: Proc. International Conference on Automatic Face and Gesture Recognition, pp. 46–53 (2000)
18. Li, S.Z., Jain, A.K. (eds.): Handbook of Face Recognition. Springer, New York (2005)
19. Li, S.Z., Zhang, Z.: FloatBoost learning and statistical face detection. IEEE Trans. Pattern Anal. Mach. Intell. **26**(9), 1112–1123 (2004)
20. Li, S.Z., Chu, R., Liao, S., Zhang, L.: Illumination invariant face recognition using near-infrared images. IEEE Trans. Pattern Anal. Mach. Intell. **29**(4), 627–639 (2007)
21. Li, S.Z., Zhao, C., Zhu, X., Lei, Z.: Learning to fuse 3D+2D based face recognition at both feature and decision levels. In: Proc. IEEE International Workshop on Analysis and Modeling of Faces and Gestures, pp. 44–54 (2005)
22. Lu, H.C., Wang, C., Chen, Y.W.: Gaze tracking by binocular vision and LBP features. In: Proc. International Conference on Pattern Recognition, pp. 1–4 (2008)
23. Lyons, M., Akamatsu, S., Kamachi, M., Gyoba, J.: Coding facial expressions with Gabor wavelets. In: Proc. IEEE International Conference on Automatic Face and Gesture Recognition, pp. 200–205 (1998)

24. Ma, B., Zhang, W., Shan, S., Chen, X., Gao, W.: Robust head pose estimation using LGBP. In: Proc. International Conference on Pattern Recognition, vol. 2, pp. 512–515 (2006)
25. Manjunath, B.S., Ohm, J.R., Vinod, V.V., Yamada, A.: Color and texture descriptors. IEEE Trans. Circuits Syst. Video Technol., 11(6), 703–715 (2001). Special Issue on MPEG-7
26. Marcel, S., Keomany, J., Rodriguez, Y.: Robust-to-illumination face localisation using active shape models and local binary patterns. Technical Report IDIAP-RR 47, IDIAP Research Institute (2006)
27. Moghaddam, B., Nastar, C., Pentland, A.: A Bayesian similarity measure for direct image matching. In: Proc. International Conference on Pattern Recognition, vol. 2, pp. 350–358 (1996)
28. Nanni, L., Lumini, A.: RegionBoost learning for 2D+3D based face recognition. Pattern Recognit. Lett. 28(15), 2063–2070 (2007)
29. Phillips, P.J., Moon, H., Rizvi, S.A., Rauss, P.J.: The FERET evaluation methodology for face-recognition algorithms. IEEE Trans. Pattern Anal. Mach. Intell. 22, 1090–1104 (2000)
30. Rodriguez, Y., Marcel, S.: Face authentication using adapted local binary pattern histograms. In: Proc. European Conference on Computer Vision, pp. 321–332 (2006)
31. Roy, A., Marcel, S.: Haar local binary pattern feature for fast illumination invariant face detection. In: Proc. British Machine Vision Conference (2009)
32. Shan, C., Gong, S., McOwan, P.W.: Conditional mutual information based boosting for facial expression recognition. In: Proc. British Machine Vision Conference (2005)
33. Shan, C., Gong, S., McOwan, P.W.: Robust facial expression recognition using local binary patterns. In: Proc. IEEE International Conference on Image Processing, pp. 370–373 (2005)
34. Shan, C., Gritti, T.: Learning discriminative LBP-histogram bins for facial expression recognition. In: Proc. British Machine Vision Conference, p. 10 (2008)
35. Shan, C.F., Gong, S.G., McOwan, P.W.: Facial expression recognition based on local binary patterns: A comprehensive study. Image Vis. Comput. 27(6), 803–816 (2009)
36. Shan, S., Zhang, W., Su, Y., Chen, X., Gao, W.: Ensemble of piecewise FDA based on spatial histograms of local (Gabor) binary patterns for face recognition. In: Proc. International Conference on Pattern Recognition, vol. 4, pp. 606–609 (2006)
37. Sun, N., Zheng, W., Sun, C., Zou, C., Zhao, L.: Gender classification based on boosting local binary pattern. In: Proc. International Symposium on Neural Networks, pp. 194–201 (2006)
38. Sun, R., Ma, Z.: Robust and efficient eye location and its state detection. In: Proc. International Symposium on Intelligence Computation and Applications, pp. 318–326 (2009)
39. Sun, Z., Tan, T., Qiu, X.: Graph matching iris image blocks with local binary pattern. In: Proc. International Conference on Biometrics, pp. 366–373 (2006)
40. Tan, X., Triggs, B.: Enhanced local texture feature sets for face recognition under difficult lighting conditions. In: Analysis and Modeling of Faces and Gestures. Lecture Notes in Computer Science, vol. 4778, pp. 168–182. Springer, Berlin (2007)
41. Tan, X., Triggs, B.: Fusing Gabor and LBP feature sets for kernel-based face recognition. In: Analysis and Modeling of Faces and Gestures. Lecture Notes in Computer Science, vol. 4778, pp. 235–249. Springer, Berlin (2007)
42. Turk, M., Pentland, A.: Eigenfaces for recognition. J. Cogn. Neurosci. 3, 71–86 (1991)
43. Varma, M., Zisserman, A.: Texture classification: Are filter banks necessary? In: Proc. IEEE Conference on Computer Vision and Pattern Recognition, vol. 2, pp. 691–698 (2003)
44. Viola, P., Jones, M.: Rapid object detection using a boosted cascade of simple features. In: Proc. IEEE Conference on Computer Vision and Pattern Recognition, pp. 511–518 (2001)
45. Wang, J.G., Yau, W.Y., Wang, H.L.: Age categorization via ECOC with fused Gabor and LBP features. In: Proc. IEEE Workshop on Applications of Computer Vision, pp. 1–6 (2009)
46. Wiskott, L., Fellous, J.-M., Kuiger, N., von der Malsburg, C.: Face recognition by elastic bunch graph matching. IEEE Trans. Pattern Anal. Mach. Intell. 19, 775–779 (1997)
47. Yan, S., Shan, S., Chen, X., Gao, W.: Locally assembled binary (LAB) feature with feature-centric cascade for fast and accurate face detection. In: Proc. IEEE Conference on Computer Vision and Pattern Recognition, pp. 1–7 (2008)
48. Yang, M.-H., Kriegman, D.J., Ahuja, N.: Detecting faces in images: a survey. IEEE Trans. Pattern Anal. Mach. Intell. 24, 34–58 (2002)

49. Yang, Z., Ai, H.: Demographic classification with local binary patterns. In: Proc. International Conference on Biometrics, pp. 464–473 (2007)
50. Zhang, G., Huang, X., Li, S.Z., Wang, Y., Wu, X.: Boosting local binary pattern LBP-based face recognition. In: Proc. Advances in Biometric Person Authentication: 5th Chinese Conference on Biometric Recognition, pp. 179–186 (2004)
51. Zhang, L., Chu, R.F., Xiang, S.M., Liao, S.C., Li, S.Z.: Face detection based on multi-block LBP representation. In: Proc. IEEE International Conference on Biometrics, pp. 11–18 (2007)
52. Zhang, W., Shan, S., Gao, W., Chen, X., Zhang, H.: Local Gabor binary pattern histogram sequence (LGBPHS): A novel non-statistical model for face representation and recognition. In: Proc. International Conference on Computer Vision, vol. 1, pp. 786–791 (2005)
53. Zhang, W., Shan, S., Qing, L., Chen, X., Gao, W.: Are Gabor phases really useless for face recognition? Pattern Anal. Appl. **12**, 301–307 (2009)
54. Zhao, W., Chellappa, R., Phillips, P.J., Rosenfeld, A.: Face recognition: A literature survey. ACM Comput. Surv. **34**(4), 399–458 (2003)

Chapter 11
Face Analysis Using Image Sequences

While many works consider moving faces only as collections of frames and apply still image based methods, recent developments indicate that excellent results can be obtained using texture based spatiotemporal representations for describing and analyzing faces in videos. Inspired by the psychophysical findings which state that facial movements can provide valuable information to face analysis, this chapter investigates the use of spatiotemporal LBP for combining facial appearance (the shape of the face) and motion (the way a person is talking and moving his/her facial features) for face, facial expression and gender recognition from videos.

11.1 Facial Expression Recognition Using Spatiotemporal LBP

This section considers the LBP based representation for dynamic texture analysis, and applies it to the problem of facial expression recognition from videos [21]. The goal of facial expression recognition is to determine the emotional state of the face, for example, happiness, sadness, surprise, neutral, anger, fear, and disgust, regardless of the identity of the face. Psychological studies [6] have shown that facial motion is fundamental to the recognition of facial expressions and humans do better job in recognizing expressions from dynamic images as opposed to mug shots.

Considering the motion of the facial region, region-concatenated descriptors on the basis of simplified VLBP are considered. Like in [3], an LBP description computed over the whole facial expression sequence encodes only the occurrences of the micro-patterns without any indication about their locations. To overcome this effect, a representation in which the face image is divided into several overlapping blocks is used. Figure 11.1 depicts overlapping 4×3 blocks with an overlap of 10 pixels. The LBP-TOP histograms in each block are computed and concatenated into a single histogram, as Fig. 11.2 shows. All features extracted from each block volume are connected to represent the appearance and motion of the facial expression sequence, as shown in Fig. 11.3. The basic VLBP features are also extracted on the basis of region motion in same way as the LBP-TOP features.

M. Pietikäinen et al., *Computer Vision Using Local Binary Patterns*,
Computational Imaging and Vision 40,
DOI 10.1007/978-0-85729-748-8_11, © Springer-Verlag London Limited 2011

Fig. 11.1 Overlapping blocks (4 × 3, overlap size = 10). (From G. Zhao and M. Pietikäinen, Dynamic texture recognition using local binary patterns with an application to facial expressions, IEEE Transactions on Pattern Analysis and Machine Intelligence, Vol. 29, Num. 6, 915–928, 2007. @2007 IEEE)

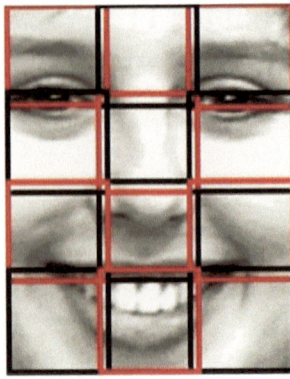

Fig. 11.2 Features in each block volume. (**a**) Block volumes; (**b**) LBP features from three orthogonal planes; (**c**) Concatenated features for one block volume with the appearance and motion

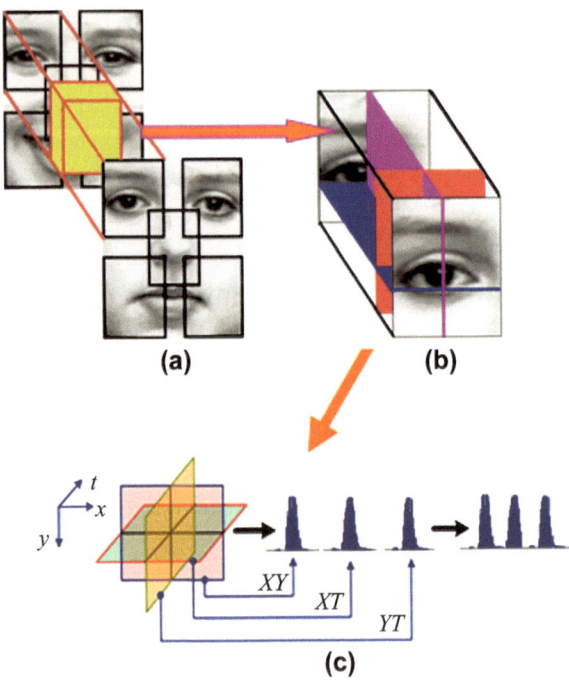

Experiment are conducted on the Cohn-Kanade database [13] which consists of 100 university students with age ranging from 18 to 30 years. Sixty-five percent were female, 15 percent African-American, and three percent Asian or Latino. Subjects were instructed by an experimenter to perform a series of 23 facial displays that included single action units and combinations of action units, six of which were based on descriptions of prototypic emotions, anger, disgust, fear, joy, sadness, and surprise. In the experiments, 374 sequences were selected from the database for basic emotional expression recognition. The selection criterion was that a sequence to be labeled is one of the six basic emotions. The sequences came from 97 subjects,

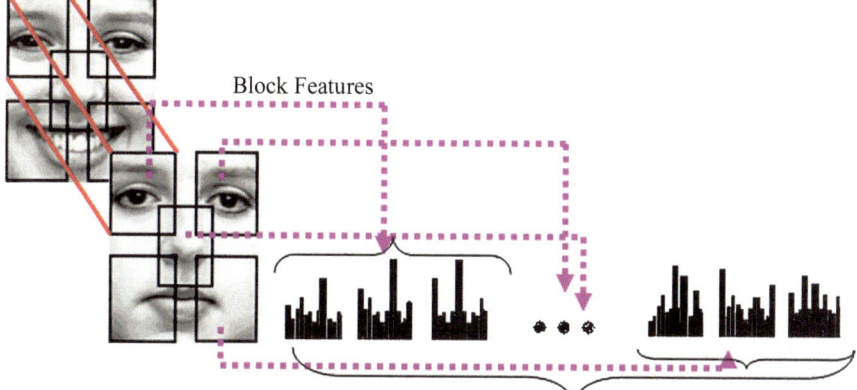

Block Features

Facial expression features from the whole sequence

Fig. 11.3 Facial expression representation

with one to six emotions per subject. Just the positions of the eyes from the first frame of each sequence were used to determine the facial area for the whole sequence. The whole sequence was used to extract the proposed LBP-TOP and VLBP features.

Figure 11.4 summarizes the confusion matrix obtained using a ten-fold cross-validation scheme on the Cohn-Kanade facial expression database. The model achieved a 96.26% overall recognition rate of facial expressions. The details of the experiments and comparison with other dynamic and static methods can be found in [21]. These experimental results clearly showed that the LBP based approach outperforms the other dynamic and static methods [4, 5, 15, 19, 20]. The approach is quite robust with respect to variations of illumination and skin color, as seen from the pictures in Fig. 11.5. It also performed well with some in-plane and out-of-plane rotated sequences. This demonstrates robustness to errors in alignment.

Spatiotemporal LBP descriptors, especially block-based LBP-TOP have been successfully utilized to many video-based applications, e.g. face recognition from videos, dynamic facial expression recognition and activity recognition. They can effectively describe appearance, horizontal motion and vertical motion from the video sequence. But block-based LBP-TOP used all the block features which makes the feature vector too long and thus the recognition cannot be done in real time. LBP-TOP was extended to include multiresolution features which are computed from different sized blocks, different neighboring samplings and different sampling scales, and utilized AdaBoost to select the slice features for all the expression classes or every class pair, to improve the performance with short feature vectors. After that, on the basis of selected slices, the location and feature types of most discriminative features for every class pair can be obtained. Figure 11.6 shows the selected features for two expression pairs. They are different and specific depending on the expressions. The detailed presentation of this work was reported in [22] and this approach was also exploited to visual speech recognition [23] (Chap. 12).

	Surprise	Happiness	Sadness	Fear	Anger	Disgust
Surprise	**98.65**		1.35			
Happiness		**96.04**		2.97		0.99
Sadness	1.37		**95.89**		2.74	
Fear		3.57	1.79	**94.64**		
Anger			3.12		**96.88**	
Disgust			2.63		2.63	**94.74**

Fig. 11.4 Confusion matrix. (From G. Zhao and M. Pietikäinen, Dynamic texture recognition using local binary patterns with an application to facial expressions, IEEE Transactions on Pattern Analysis and Machine Intelligence, Vol. 29, Num. 6, 915–928, 2007. @2007 IEEE)

Fig. 11.5 Variation of illumination. (From G. Zhao and M. Pietikäinen, Dynamic texture recognition using local binary patterns with an application to facial expressions, IEEE Transactions on Pattern Analysis and Machine Intelligence, Vol. 29, Num. 6, 915–928, 2007. @2007 IEEE)

Most of the current facial expression recognition studies are carried out in visible light spectrum. But visible light usually changes with locations, and can also vary with time, which can cause significant variations in image appearance and texture. Dynamic facial expression recognition from near-infrared (NIR) video sequences was investigated in [18]. Experiments on a new Oulu-CASIA NIR database showed promising and robust results against illumination variations. Figure 11.7 demonstrates results taken from an example image sequence captured with an NIR web camera, showing the detected faces and classified expressions.

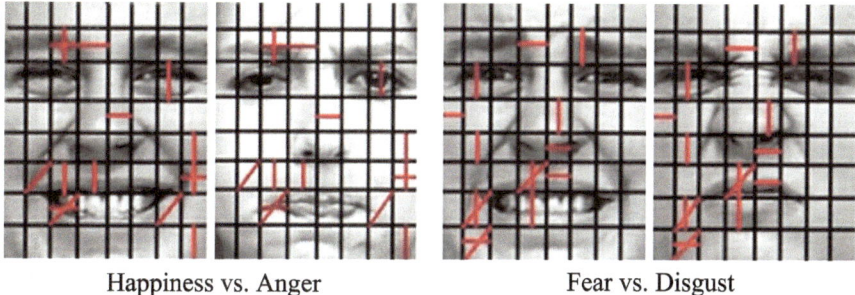

|Happiness vs. Anger|Fear vs. Disgust|

Fig. 11.6 Selected 15 slices for expression Happiness vs. Anger (*left*) and Fear vs. Disgust (*right*)

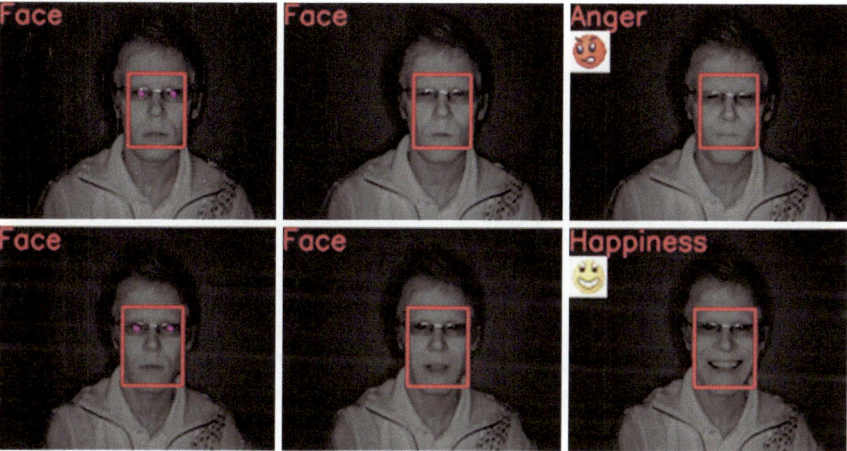

Fig. 11.7 Facial expression recognition from a NIR web camera. From left to right are the start, middle and the end of the sequences

11.2 Face Recognition from Videos

Despite the evidences from psychophysics and neuroscience which indicate that facial movements can provide valuable information to face recognition, most automatic systems dealing with face analysis in video sequences use only the static information as it is unclear how the dynamic cue can be integrated and exploited. Thus, most research has limited the scope of the problem by applying methods developed for still images to some selected frames, for instance. Only recently have researchers started to truly address the problem of face recognition from video sequences using spatiotemporal representations.

Unfortunately, the few methods which use spatiotemporal representations for face recognition suffer from at least one of the following drawbacks: (i) the local information which is shown to be important to facial image analysis [12] is not well exploited with holistic methods such as HMMs (Hidden Markov models); (ii) while only personal specific facial dynamics are useful for discriminating between different persons, the intra-personal temporal information which is related to facial expression and emotions is also encoded and used; and (iii) equal weights are given to the spatiotemporal features despite the fact that some of the features contribute to recognition more than others. To overcome these limitations, an effective approach for face recognition from videos that uses local volumetric spatiotemporal LBP features and selects only the useful facial dynamics needed for recognition was developed in [10, 11]. The idea consists of looking at a face sequence as a selected set of volumes (or rectangular prisms) from which local histograms of extended volume local binary pattern code occurrences (EVLBP) are extracted. Thus, each face sequence is first divided into several overlapping rectangular prisms of different sizes, from which local spatiotemporal LBP histograms are extracted. Then, instead of simply concatenating the local histograms into a single histogram, AdaBoost learning algorithm is used for automatically determining the optimal size and locations of the local rectangular prisms, and more importantly for selecting the most discriminative patterns for recognition while discarding the features which may hinder the recognition process.

To tackle the problem of selecting only the spatiotemporal information which is useful for recognition while discarding the information related to facial expressions and emotions, AdaBoost learning technique [8] which has shown its efficiency in feature selection tasks is adopted. The idea is to separate the facial information into intra and extra classes, and then use only the extra-class features for recognition.

First, the training face sequences are segmented into several overlapping shots of F frames each in order to increase the number of training data. Then, all combinations of face sequence pairs for the intra and extra classes are considered. For each pair $(sequence_i^1, sequence_i^2)$, both face sequences are scanned with rectangular prisms of different sizes. At each stage, histograms of spatiotemporal LBP code are extracted from the local rectangular prisms and the χ^2 (chi-square) distances are computed between the two local histograms.

Thus, for each pair of face sequences, a feature vector X_i whose elements are χ^2 distances is obtained. Let $Y_i \in \{+1, -1\}$ denote the class label of X_i where $Y_i = +1$ if the pair $(sequence_i^1, sequence_i^2)$ defines an extra-class pair (i.e. the two sequences are from different persons) and $Y_i = -1$ otherwise. This results in a set of training samples $\{(X_1, Y_1), (X_2, Y_2), \ldots, (X_N, Y_N)\}$. Details on the procedure of constructing the training data can be found in [11] .

Given the constructed training sets, AdaBoost learning algorithm [8] is applied in order to (i) select a subset of rectangular prisms from which spatiotemporal LBP features should be computed, and (ii) learn and determine the weights of these selected features.

Once the rectangle prisms are selected and their weights are determined, the recognition of a given probe video sequence is performed by extracting local histograms of spatiotemporal LBP patterns from the selected prisms and then applying nearest neighbor classification using weighted χ^2 distance.

For experimental analysis, three different publicly available video face databases (MoBo [9], Honda/UCSD [14] and CRIM [7]) are considered in order to ensure an extensive evaluation of the proposed approach and the benchmark methods against changes caused by different factors including face image resolution, illumination variations, head movements, facial expressions and the size of the database. For comparison, the implementation of five different algorithms including Hidden Markov models (HMMs) [16] and Auto-Regressive and Moving Average (ARMA) models [1] as benchmark methods for spatiotemporal representations, and PCA, LDA and LBP [3] for still image based ones were also considered.

The experimental analysis showed that, in some cases, the methods which use only the facial structure (such as PCA, LDA and the original LBP) can outperform the spatiotemporal approaches. This can be explained by the fact that some facial dynamics is not useful for recognition. In other terms, this means that some part of the temporal information is useful for recognition while another part may also hinder the recognition. Obviously, the useful part defines the extra-personal characteristics while the non-useful one concerns the intra-class information such as facial expressions and emotions. For recognition, one should then select only the extra-personal characteristics. The proposed approach tackles the problem of selecting only the spatiotemporal information which is useful for recognition using AdaBoost learning technique. It classifies the facial information into intra and extra classes, and then uses only the extra-class spatiotemporal LBP features for recognition. In comparison to the benchmark methods, the approach yielded in a significant increase in the recognition rates. The significant increases in the recognition rates can be explained by the following: (i) the LBP based spatiotemporal representation, in contrast to the HMM based approach, is very efficient as it codifies the local facial dynamics and structure, (ii) the temporal information extracted by the spatiotemporal LBP features consists of both intra and extra personal dynamics (facial expression and identity). Therefore, there was need for performing feature selection. This yielded in excellent results.

Analyzing the selected local regions (the rectangular prisms) from which the spatiotemporal LBP features were collected, it was noticed that the dynamics of the whole face and the eye area are more important than that of the mouth region for identity recognition. This is a little surprising in the sense that one can expect that the mouth region would play an important role as it is the most non-rigid region of the face when an individual is talking. Perhaps, mouth region does play an important role but for facial expression recognition. Figure 11.8 shows examples of the most discriminative spatiotemporal regions returned by AdaBoost for CRIM face sequences and from which spatiotemporal LBP features are extracted. Notice that these four first selected features are extracted from global and local regions. This supports the results of other researchers indicating that both global and local features are useful for recognition. Table 11.1 summarizes the obtained results using the different methods (PCA, LDA, LBP, HMM, ARMA, VLBP and EVLBP) on the three databases (MoBo, Honda/UCSD and CRIM). Details of these experiments can be found in [10, 11].

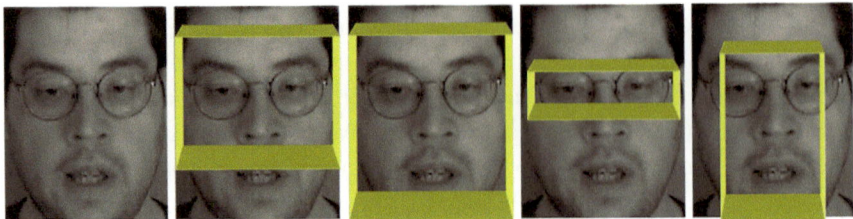

Fig. 11.8 Examples of the four first selected rectangular prisms from which spatiotemporal LBP features are extracted on CRIM face sequences

Table 11.1 Summary of the obtained results using the different methods on the three databases

Method	Results on MoBo	Results on Honda/UCSD	Results on CRIM
PCA	87.1%	69.9%	89.7%
LDA	90.8%	74.5%	91.5%
LBP [2]	91.3%	79.6%	93.0%
HMM [16]	92.3%	84.2%	85.4%
ARMA [1]	93.4%	84.9%	80.0%
VLBP [21]	90.3%	78.3%	88.7%
VLBP+AdaBoost [11]	96.5%	89.1%	94.4%
EVLBP+AdaBoost [11]	**97.9%**	**96.0%**	**98.5%**

11.3 Gender Classification from Videos

This section describes another challenging problem which is spatiotemporal based gender recognition from videos. The aim of gender recognition is to determine whether the person whose face is in the given video is a man or a woman. Determining such information is useful for many applications such as more affective Human Computer Interaction (HCI), restricting access to certain areas based on gender, collecting demographic information in public places, counting the number of women entering a retail store and so on.

Similarly to the face recognition experiments, the LBP methodology (as described in the previous section) was adopted to built a system for gender recognition from video using spatiotemporal LBP features and AdaBoost learning. The main difference lies in the preparation of the training data as the system is trained to select spatiotemporal LBP features (i.e, combination of appearance and facial dynamics) which discriminate best between men and women classes.

For the experimental analysis, three different publicly available video face databases (namely CRIM [7], VidTIMIT [17] and Cohn-Kanade [13]) were considered. The datasets contain a balanced number of male's and female's sequences and include several subjects moving their facial features by uttering phrases, reading broadcast news or expressing emotions. The datasets were randomly segmented to extract over 4 000 video shots of 15 to 300 frames each. From each shot or sequence,

Table 11.2 Gender classification results on test videos of familiar (*columns 1–3*) and unfamiliar subjects (*columns 4–6*). The methods are based on appearance only (*1st and 2nd rows*), and combination of appearance and motion using spatiotemporal LBP and AdaBoost (*3rd row*)

Method	Gender classification rate					
	Subjects seen during training			Subjects unseen during training		
	20×20	40×40	60×60	20×20	40×40	60×60
Pixels+SVM	93.1%	93.3%	91.9%	88.5%	89.4%	88.2%
LBP+SVM	94.0%	94.4%	95.4%	**90.1%**	**90.6%**	**91.0%**
Spatiotemporal LBP+AdaBoost	**100%**	**100%**	**100%**	79.2%	81.5%	78.6%

the eye positions were automatically detected from the first frame. The determined eye positions are then used to crop the facial area in the whole sequence. Finally the resulted images are scaled into three different resolutions: 20×20, 40×40 and 60×60 pixels.

For evaluation, 5-fold cross validation test scheme was adopted by dividing the 4 000 sequences into five groups and using the data from four groups for training and the left group for testing. This process was repeated five times and the average classification rates are reported. When dividing the data into training and test sets, two scenarios were explicitly considered. In the first one, a same person may appear in both training and test sets with face sequences completely different in the two sets due to facial expression, lighting, facial pose etc. The goal of this scenario is to analyze the performance of the methods in determining the gender of familiar persons seen under different conditions. In the second scenario, the test set consists only of persons who are not included in the training sets. This is equivalent to train the system on one or more databases and then do evaluation on other (different) databases. The goal of this scenario is to test the generalization ability of the methods to determine the gender of unseen persons.

For comparison, two benchmark methods based only on still images were also implemented. The first method uses normalized raw pixel values as inputs to an SVM classifier while the second method uses LBP features with SVM. All built SVM classifiers are using second degree polynomial kernel functions. In both methods, each frame is processed separately and then the results are combined through majority voting which means identifying the gender in every frame and then fusing the results.

The gender classification results using the proposed approach (spatiotemporal LBP + AdaBoost) and the two benchmark methods (Pixels+SVM and LBP+SVM) in both scenarios (familiar and unfamiliar) and three different image resolutions are summarized in Table 11.2. One can notice that all methods gave better results with familiar faces than unfamiliar ones. This is not surprising and can be explained by the fact that probably the methods did not rely only on gender features for classification but may also exploited information about face identity. For familiar faces, the combination of appearance and motion using the proposed approach yielded in perfect classification rates of 100%. This proves that the system succeeded in learning and recognizing the facial behaviors of the subjects even under different conditions

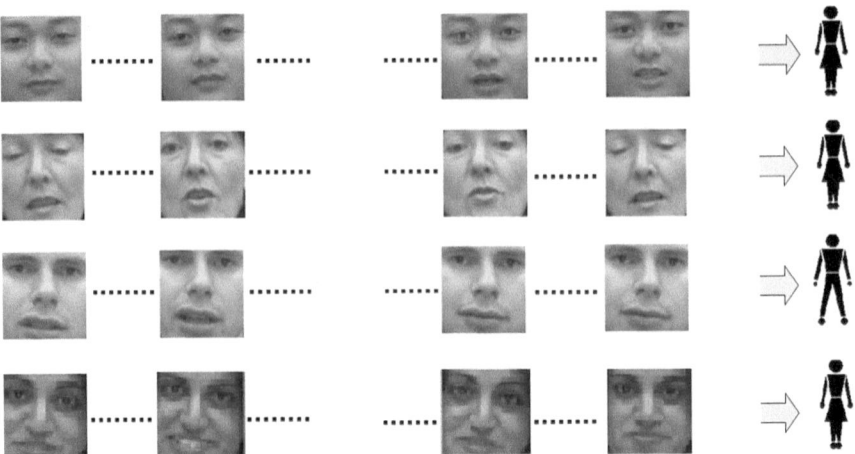

Fig. 11.9 Example of gender classification results using boosted spatiotemporal LBP features

of facial expression, lighting and facial pose. For unfamiliar faces, the combination of appearance and motion using the proposed approach yielded in classification rate of about 80% which is not so bad. The best results for unfamiliar faces are obtained using only still images without motion (using LBP+SVM approach). This may indicate that incorporating motion information with appearance was useful for only familiar faces but not with unfamiliar ones. The results also show that image resolution does not affect very much gender classification performance and this confirms the conclusions of many other researchers. Some examples of gender recognition results using the proposed approach with unfamiliar faces are shown in Fig. 11.9. The first row shows an example in which the system failed to classify a video of a male. In the second, third and forth rows, the system successfully determined the gender of the person.

Analyzing the misclassification errors made by the different methods, it was noticed that female's sequences are harder to classify than male's ones. In average, about 65% of misclassifications concerned female's sequences which are classified as males. Perhaps, this could be explained by the fact that when only facial areas are used for gender classification, the presence of moustaches and beards helps more the classification of male's images. However, one can expect better classification of female's images when external features such as hair are also included.

11.4 Discussion

Because finding an efficient spatiotemporal representation for face analysis from videos is challenging, most of the existing works limit the scope of the problem by discarding the facial dynamics and only considering the structure. Motivated by the psychophysical findings which indicate that facial movements can provide valuable information to face analysis, spatiotemporal LBP approaches for face, facial expression and gender recognition from videos were described. The extensive experimental analysis clearly assessed the excellent performance of the LBP based spatiotemporal representations for describing and analyzing faces in videos. The efficiency of the proposed approaches can be explained by the local nature of the spatiotemporal LBP descriptions, combined with the use of boosting for selecting the optimal features.

References

1. Aggarwal, G., Chowdhury, A.R., Chellappa, R.: A system identification approach for video-based face recognition. In: Proc. International Conference on Pattern Recognition, vol. 4, pp. 175–178 (2004)
2. Ahonen, T., Hadid, A., Pietikäinen, M.: Face recognition with local binary patterns. In: European Conference on Computer Vision. Lecture Notes in Computer Science, vol. 3021, pp. 469–481. Springer, Berlin (2004)
3. Ahonen, T., Hadid, A., Pietikäinen, M.: Face description with local binary patterns: Application to face recognition. IEEE Trans. Pattern Anal. Mach. Intell. 28(12), 2037–2041 (2006)
4. Aleksic, S.P., Katsaggelos, K.A.: Automatic facial expression recognition using facial animation parameters and multi-stream HMMs. IEEE Trans. Inf. Forensics Secur. 1(1), 3–11 (2006)
5. Bartlett, M.S., Littlewort, G., Fasel, I., Movellan, R.: Real time face detection and facial expression recognition: Development and application to human computer interaction. In: Proc. CVPR Workshop on Computer Vision and Pattern Recognition for Human-Computer Interaction, vol. 5, p. 53 (2003)
6. Bassili, J.: Emotion recognition: The role of facial movement and the relative importance of upper and. J. Pers. Soc. Psychol. 37, 2049–2059 (1979)
7. CRIM: http://www.crim.ca/
8. Freund, Y., Schapire, R.: A decision-theoretic generalization of on-line learning and an application to boosting. J. Comput. Syst. Sci. 55(1), 119–139 (1997)
9. Gross, R., Shi, J.: The CMU Motion of Body (MoBo) database. Technical Report CMU-RI-TR-01-18, Robotics Institute, CMU (2001)
10. Hadid, A., Pietikäinen, M.: Combining appearance and motion for face and gender recognition from videos. Pattern Recognit. 42(11), 2818–2827 (2009)
11. Hadid, A., Pietikäinen, M., Li, S.Z.: Learning personal specific facial dynamics for face recognition from videos. In: Analysis and Modeling of Faces and Gestures. Lecture Notes in Computer Science, vol. 4778, pp. 1–15. Springer, Berlin (2007)
12. Heisele, B., Ho, P., Wu, J., Poggio, T.: Face recognition: Component based versus global approaches. Comput. Vis. Image Underst. 91(1–2), 6–21 (2003)
13. Kanade, T., Cohn, J.F., Tian, Y.: Comprehensive database for facial expression analysis. In: Proc. International Conference on Automatic Face and Gesture Recognition, pp. 46–53 (2000)
14. Lee, K.C., Ho, J., Yang, M.H., Kriegman, D.: Video-based face recognition using probabilistic appearance manifolds. In: Proc. IEEE Conference on Computer Vision and Pattern Recognition, pp. 313–320 (2003)
15. Littlewort, G., Bartlett, M., Fasel, I., Susskind, J., Movellan, J.: Dynamics of facial expression extracted automatically from video. In: Proc. IEEE Workshop on Face Processing in Video, pp. 80–80 (2004)
16. Liu, X., Chen, T.: Video-based face recognition using adaptive hidden Markov models. In: Proc. IEEE Conference on Computer Vision and Pattern Recognition, pp. 340–345 (2003)
17. Sanderson, C., Paliwal, K.K.: Noise compensation in a person verification system using face and multiple speech feature. Pattern Recognit. 36(2), 293–302 (2003)
18. Taini, M., Zhao, G., Pietikäinen, M.: Facial expression recognition from near-infrared video sequences. In: Proc. International Conference on Pattern Recognition, pp. 1–4 (2008)

19. Tian, Y.: Evaluation of face resolution for expression analysis. In: Proc. Computer Vision and Pattern Recognition Workshop, p. 7 (2004)
20. Yeasin, M., Bullot, B., Sharma, R.: From facial expression to level of interest: A spatio-temporal approach. In: Proc. IEEE Conference on Computer Vision and Pattern Recognition, pp. 922–927 (2004)
21. Zhao, G., Pietikäinen, M.: Dynamic texture recognition using local binary patterns with an application to facial expressions. IEEE Trans. Pattern Anal. Mach. Intell. **29**(6), 915–928 (2007)
22. Zhao, G., Pietikäinen, M.: Boosted multi-resolution spatiotemporal descriptors for facial expression recognition. Pattern Recognit. Lett. **30**(12), 1117–1127 (2009)
23. Zhao, G., Barnard, M., Pietikäinen, M.: Lipreading with local spatiotemporal descriptors. IEEE Trans. Multimed. **11**(7), 1254–1265 (2009)

Chapter 12
Visual Recognition of Spoken Phrases

Visual speech information plays an important role in speech recognition under noisy conditions or for listeners with hearing impairment. In this chapter, local spatiotemporal descriptors are utilized to represent and recognize spoken isolated phrases based solely on visual input [14]. Positions of the eyes are used for localizing the mouth regions in face images and then spatiotemporal local binary patterns extracted from these regions are used for describing phrase sequences. Experiments show promising results. Advantages of the approach include local processing and robustness to monotonic gray-scale changes. Moreover, no error prone segmentation of moving lips is needed.

12.1 Related Work

It is well known that human speech perception is a multimodal process. Visual observation of the lips, teeth and tongue offers important information about the place of pronunciation articulation. In some researches, lipreading combined with face and voice is studied to help biometric identification [3, 4, 6]. There is also much work focusing on audio-visual speech recognition (AVSR) [1, 5, 9], trying to find effective ways of combining visual information with existing audio-only speech recognition systems (ASR). McGurk effect [8] demonstrates that inconsistency between audio and visual information can result in perceptual confusion. Visual information plays an important role especially in noisy environments or for the listeners with hearing impairment. A human listener can use visual cues, such as lip and tongue movements, to enhance the level of speech understanding. The process of using visual modality is often referred to as lipreading which is to make sense of what someone is saying by watching the movement of his lips.

Comprehensive reviews of automatic audio-visual speech recognition can be found in [9, 10]. Extraction of a discriminative set of visual observation vectors is the key element of an AVSR system. Geometric features, appearance features and combined features are commonly used for representing visual information.

M. Pietikäinen et al., *Computer Vision Using Local Binary Patterns*, Computational Imaging and Vision 40, DOI 10.1007/978-0-85729-748-8_12, © Springer-Verlag London Limited 2011

Most of the researches focus on using visual information to improve speech recognition. Audio features are still the main part and play more important role. However, in some cases, it is difficult to extract useful information from the audio. There are many applications in which it is necessary to recognize speech under extremely adverse acoustic environments. Detecting a person's speech from a distance or through a glass window, understanding a person speaking among a very noisy crowd of people, and monitoring a speech over TV broadcast when the audio link is weak or corrupted, are some examples. Furthermore, for the persons with hearing impairment, visual information is the only source of information from TV broadcast or speeches if there is no assisting sign language. In these applications, the performance of traditional speech recognition is very limited. There are a few works focusing on the lip movement representations for speech recognition solely with visual information [2, 7, 11, 12]. So addressing this problem could improve the quality of human-computer interaction (HCI).

It appears that most of the research on visual speech recognition based on the appearance features has considered global features of lip or mouth images, but omitting the local features. Local features can describe the local changes of images in space and time. In this chapter, the recognition of isolated phrases using only visual information is considered. An appearance feature representation based on spatiotemporal local binary patterns is utilized, taking into account the motion of mouth region and time order in pronunciation. A Support Vector Machine (SVM) classifier is utilized for recognition. The details of the methods and experimental results can be seen in [14].

12.2 System Overview

The system consists of three stages, as shown in Fig. 12.1. The first stage is a combination of discriminative classifiers that first detects the face, and then the eyes. The positions of the eyes are used to localize the mouth region. The second stage extracts the visual features from the mouth movement sequence. The role of the last stage is to recognize the input utterance using SVM classifier.

12.3 Local Spatiotemporal Descriptors for Visual Information

Due to its ability to describe spatiotemporal signals, robustness to monotonic grayscale changes caused e.g. by illumination variations, the LBP-TOP is utilized to represent the mouth movements. Considering the motion of the mouth region, the descriptors are obtained by concatenating local binary patterns on three orthogonal planes from the utterance sequence: XY, XT and YT, considering only the co-occurrence statistics in these three directions. Figure 12.2(a) demonstrates the volume of utterance sequence, (b) shows image in the XY plane, (c) is image in XT plane providing visual impression of one row changing in time, while (d) describes

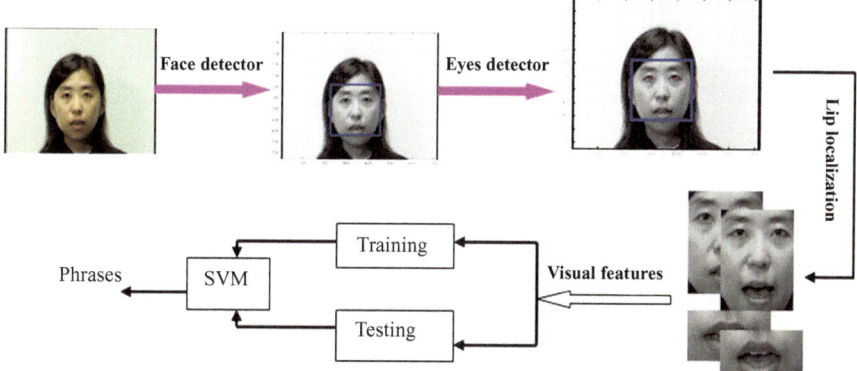

Fig. 12.1 System diagram

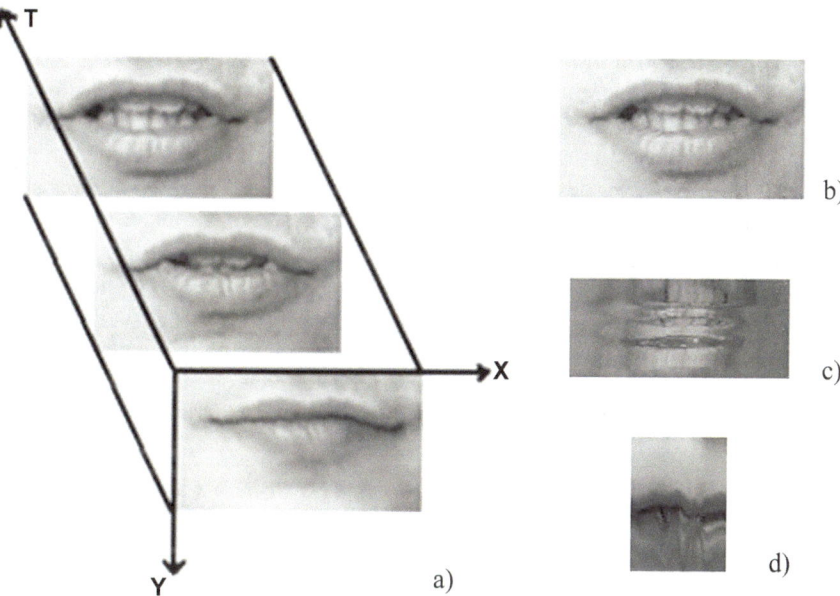

Fig. 12.2 (**a**) Volume of utterance sequence. (**b**) Image in XY plane (147×81). (**c**) Image in XT plane (147×38) in $y = 40$ (last row is pixels of $y = 40$ in first image). (**d**) Image in TY plane (38×81) in $x = 70$ (first column is the pixels of $x = 70$ in first frame)

the motion of one column in temporal space. An LBP description computed over the whole utterance sequence encodes only the occurrences of the micro-patterns without any indication about their locations. To overcome this effect, a representation which consists of dividing the mouth image into several overlapping blocks is used.

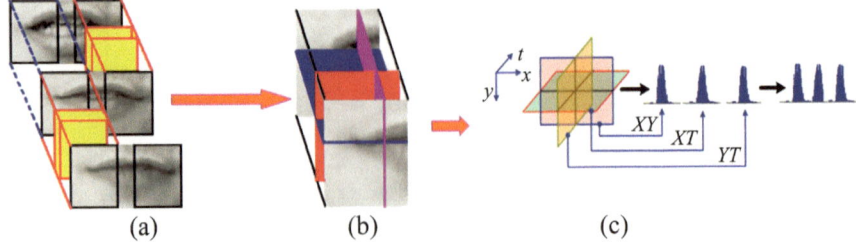

Fig. 12.3 Features in each block volume. (**a**) Block volumes. (**b**) LBP features from three orthogonal planes. (**c**) Concatenated features for one block volume with the appearance and motion

However, taking only into account the locations of micro-patterns is not enough. When a person utters a command phrase, the words are pronounced in order, for instance "you-see" or "see-you". If the time order is not considered, these two phrases would get almost the same features. To overcome this effect, the whole sequence is not only divided into block volumes according to spatial regions but also in time order, as Fig. 12.3(a) shows.

The LBP-TOP histograms in each block volume are computed and concatenated into a single histogram, as Fig. 12.3 shows. All features extracted from each block volume are connected to represent the appearance and motion of the mouth region sequence. In this way, a description of the phrase utterance on three different levels of locality is obtained. The labels (bins) in the histogram contain information from three orthogonal planes, describing appearance and temporal information at the pixel level. The labels are summed over a small block to produce information on a regional level expressing the characteristics for the appearance and motion in specific locations and time segment, and all information from the regional level is concatenated to build a global description of the mouth region motion.

A histogram of the mouth movements can be defined as

$$H_{r,c,d,j,i} = \sum_{x,y,t} I\left\{ f_j(x,y,t) = i \right\}, \quad i = 0, \cdots, n_j - 1; j = 0, 1, 2 \quad (12.1)$$

in which n_j is the number of different labels produced by the LBP operator in the jth plane ($j = 0$: XY, 1: XT and 2: YT), $f_j(x,y,t)$ expresses the LBP code of central pixel (x,y,t) in the jth plane, r is the index of rows, c is of columns and d is of time of block volume.

$$I\{A\} = \begin{cases} 1, & \text{if } A \text{ is true;} \\ 0, & \text{if } A \text{ is false.} \end{cases} \quad (12.2)$$

The histograms must be normalized to get a coherent description:

$$N_{r,c,d,j,i} = \frac{H_{r,c,d,j,i}}{\sum_{k=0}^{n_j-1} H_{r,c,d,j,k}}. \quad (12.3)$$

Table 12.1 Phrases included in the dataset

C1 "Excuse me"	C6 "See you"
C2 "Goodbye"	C7 "I am sorry"
C3 "Hello"	C8 "Thank you"
C4 "How are you"	C9 "Have a good time"
C5 "Nice to meet you"	C10 "You are welcome"

12.4 Experiments

12.4.1 Dataset Description

A visual speech dataset-OuluVS database (http://www.cse.oulu.fi/MVG/Downloads) was collected for performance evaluation.

A SONY DSR-200AP 3CCD-camera with frame rate 25 fps was used to collect the data. The image resolution was 720 by 576 pixels. The dataset includes twenty persons, each uttering ten everyday's greetings one to five times. These short phrases are listed in Table 12.1.

The subjects were asked to sit on a chair. The distance between the speaker and the camera was about 160 cm. He/she was then asked to read ten phrases which were written on a paper, each phrase one to five times. The data collection was done in two parts: at first from ten persons and four days later from the ten remaining ones. Seventeen males and three females are included, nine of whom wear glasses. Speakers are from four different countries, so they have different pronunciation habits including different speeds.

Totally, 817 sequences from 20 speakers were used in the experiments.

12.4.2 Experimental Results

Because the face images in the database are of good quality and almost all of them are frontal faces, detection of faces and eyes is quite easy. The positions of the two eyes in the first frame of each sequence were given by the eye detector automatically and then these positions were used to determine the fine facial area and localize the mouth region using pre-defined ratio parameters for the whole sequence.

When extracting the local patterns, not only locations of micro-patterns but also the time order in articulation were taken into account, so the whole sequence is divided into block volumes according to not only spatial regions but also time order.

In the recognition, an SVM classifier was selected since it is well founded in statistical learning theory and has been successfully applied to various object detection tasks in computer vision. Since SVM is only used for separating two sets of points, the 10-phrase classification problem is decomposed into 45 two-class problems ("Hello"–"Excuse Me", "I am sorry"–"Thank you", "You are welcome"–"Have a good time", etc.), then a voting scheme is used to accomplish recognition.

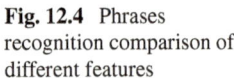

Fig. 12.4 Phrases
recognition comparison of
different features

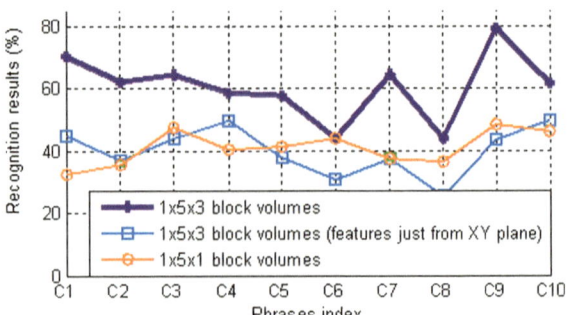

Here, after the comparison of linear, polynomial and RBF kernels in experiments, the second degree polynomial kernel function is used, because it provided the best results. Sometimes more than one class gets the highest number of votes. In this case, 1-NN template matching is applied to these classes to reach the final result. This means that in training, the spatiotemporal LBP histograms of utterance sequences belonging to a given class are averaged to generate a histogram template for that class. In recognition, a nearest-neighbor classifier is adopted.

According to tests, parameter values $P_{XY} = P_{XT} = P_{YT} = 8$, $R_X = R_Y = R_T = 3$ and an overlap ratio of 70% of the original non-overlapping block size were selected. After experimenting with different block sizes, it was chosen to use $1 \times 5 \times 3$ (rows by columns by time segments) blocks in the experiments.

For the speaker-independent experiments, leave-one-speaker-out is utilized in the testing procedure. In each run training was done on 19 speakers in the data set, while testing was performed on the remaining one. The same procedure was repeated for each speaker and the overall results were obtained using M/N (M is the total number of correctly recognized sequences and N is the total number of testing sequences).

Figure 12.4 shows the recognition results using three different features. As expected, the result of the features from three planes is better than that just from the appearance (XY) plane which justifies the effectiveness of the feature combining appearance with motion. The features with $1 \times 5 \times 1$ block volumes omitted the pronunciation order, providing a lower performance than those with $1 \times 5 \times 3$ block volumes for almost all the tested phrases. It can be seen from Fig. 12.4 that the recognition rates of phrases "See you" (C6) and "Thank you"(C8) are lower than others because the utterances of these two phrases are quite similar, just different in the tongue's position. If taking those two phrases as one class, the recognition rate would be 4% higher.

For speaker-dependent experiments, the leave-one utterance-out is utilized for cross validation because there are not abundant samples for each phrase of each speaker. Totally ten speakers with at least three training samples for each phrase are selected for this experiment, because too few training samples, for instance, one or two, could bias the recognition rate. In the experiments, every utterance is left out, and the rest utterances are trained for every speaker. Figure 12.5 presents a detailed comparison of the results for every subject.

Fig. 12.5 Speaker-dependent recognition results for every subject

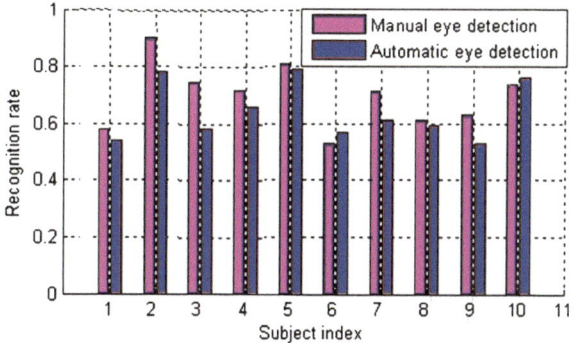

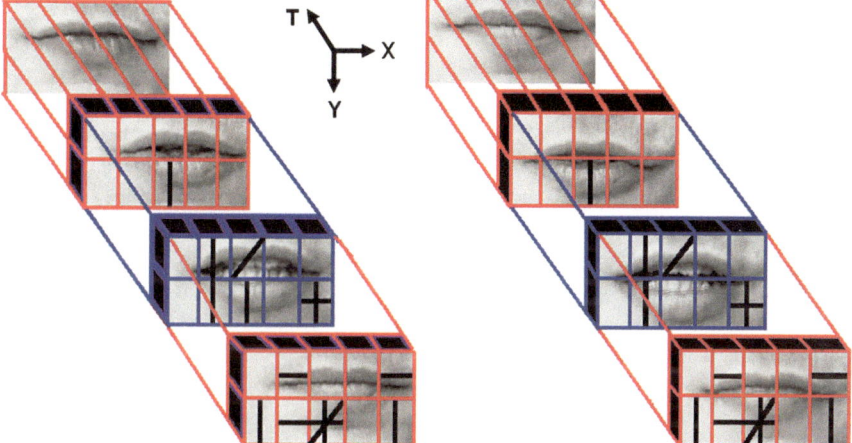

Fig. 12.6 Selected 15 slices for phrases "See you" and "Thank you". "|" in the blocks means the YT slice (vertical motion) is selected, and "−" the XT slice (horizontal motion), "/" means the appearance XY slice

12.4.3 Boosting Slice Features

A feature selection approach using AdaBoost was also investigated to select more important slices (principal appearance and motion). This approach was presented in [13] to improve the lipreading performance [14].

Figure 12.6 shows the selected slices for similar phrases "see you" and "thank you". These phrases were the most difficult to recognize because they are quite similar in the latter part containing the same word "you". The selected slices are mainly in the first and second part of the phrase, just one vertical slice is from the last part. The selected features are consistent with the human intuition. The phrases "excuse me" and "I am sorry" are different throughout the whole utterance, and the selected features also come from the whole pronunciation as shown in Fig. 12.7. With the feature selection strategy, more specific and adaptive features are selected

Fig. 12.7 Selected 15 slices for phrases "Excuse me" and "I am sorry". "|" in the blocks means the YT slice is selected, and "−" the XT slice, "/" means the appearance XY slice

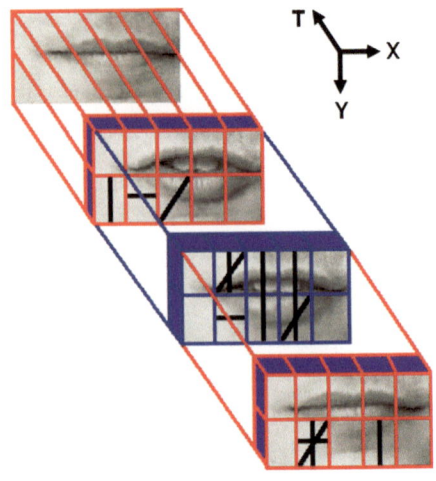

for different pairs of phrase classes, as shown in Figs. 12.6 and 12.7, providing more discriminative features.

12.5 Discussion

In this chapter, local spatiotemporal descriptor is utilized for visual speech recognition, considering the spatial region and pronunciation order in the utterance. The movements of mouth regions are described using local binary patterns from XY, XT and YT planes, combining local features from pixel, block and volume levels. Reliable lip segmentation and tracking is a major problem in automatic visual speech recognition, especially in poor imaging conditions. The presented approach avoids this using local spatiotemporal descriptors computed from mouth regions which are much easier to extract than lips. With this approach no error prone segmentation of moving lips is needed. Experiments on a dataset collected from 20 persons showed very promising results. For ten spoken phrases the obtained speaker-independent recognition rate is 62% and speaker-dependent result around 70%.

Multiresolution features and feature selection approach are presented and the preliminary experimental results show the effectiveness of selecting principal appearance and motion for specific class pairs.

Compared with the state-of-the-art, this method does not need to (1) segment lip contours; (2) track lips in the subsequent frames; (3) select constant illumination or perform illumination correction; (4) align lip features with respect to the canonical template or normalize the mouth images to a fixed size as done by most of the papers.

Recently, a video-normalization scheme [16] was presented to improve the lipreading using LBP-TOP features. Firstly, a simple deterministic model was proposed to seek a low-dimensional manifold where visual features extracted from the frames of a video can be projected onto a continuous deterministic curve embedded in a path graph. Moreover, it can map arbitrary points on the curve back into the image space, making it suitable for temporal interpolation. With this preprocessing, temporally interpolated videos are suitable for extracting multi-resolution features and thus the speaker independent lipreading performance got significantly improved.

Continuous speech segmentation and classification are also attracting lots of attention, e.g. using viseme models to improve the quality of lipreading. Moreover, it is also of interest to combine visual and audio information to promote speech recognition, and to apply this methodology to speaker identification using lipreading [15] and human-robot interaction in a smart environment.

References

1. Arsic, I., Thiran, J.P.: Mutual information eigenlips for audio-visual speech. In: Proc. European Signal Processing Conference (2006)
2. Chiou, G.I., Hwang, J.N.: Lipreading from color video. IEEE Trans. Image Process. **6**(8), 1192–1195 (1997)
3. Fox, N., Gross, R., Chazal, P.: Person identification using automatic integration of speech, lip and face experts. In: Proc. ACM SIGMM Workshop on Biometrics Methods and Applications, pp. 25–32 (2003)
4. Frischholz, R.W., Dieckmann, U.: BioID: A multimodal biometric identification system. Computer **33**(2), 64–68 (2000)
5. Gurban, M., Thiran, J.P.: Audio-visual speech recognition with a hybrid SVM-HMM system. In: Proc. European Signal Processing Conference, p. 4 (2005)
6. Luettin, J., Thacher, N.A., Beet, S.W.: Speaker identification by lipreading. In: Proc. International Conference on Spoken Language Proceedings, pp. 62–64 (1996)
7. Matthews, I., Cootes, T.F., Bangham, J.A., Cox, S., Harvey, R.: Extraction of visual features for lipreading. IEEE Trans. Pattern Anal. Mach. Intell. **24**(2), 198–213 (2002)
8. McGurk, H., MacDonald, J.: Hearing lips and seeing voices. Nature **264**, 746–748 (1976)
9. Potamianos, G., Neti, C., Gravier, G., Garg, A., Senior, A.: Recent advances in the automatic recognition of audio-visual speech. Proc. IEEE **91**(9), 1306–1326 (2003)
10. Potamianos, G., Neti, C., Luettin, J., Matthews, I.: Audio-visual automatic speech recognition: an overview. In: Issues in Visual and Audio-Visual Speech Processing. MIT Press, Cambridge (2004)
11. Saenko, K., Livescu, K., Glass, J., Darrell, T.: Production domain modeling of pronunciation for visual speech recognition. In: Proc. International Conference on Acoustics, Speech, and Signal Processing, vol. 5, pp. 473–476 (2005)
12. Saenko, K., Livescu, K., Siracusa, M., Wilson, K., Glass, J., Darrell, T.: Visual speech recognition with loosely synchronized feature streams. In: Proc. International Conference on Computer Vision, pp. 1424–1431 (2005)
13. Zhao, G., Pietikäinen, M.: Boosted multi-resolution spatiotemporal descriptors for facial expression recognition. Pattern Recognit. Lett. **30**(12), 1117–1127 (2009)
14. Zhao, G., Barnard, M., Pietikäinen, M.: Lipreading with local spatiotemporal descriptors. IEEE Trans. Multimed. **11**(7), 1254–1265 (2009)
15. Zhao, G., Huang, X., Gizatdinova, Y., Pietikäinen, M.: Combining dynamic texture and structural features for speaker identification. In: Proc. ACM Multimedia Workshop on Multimedia in Forensics, Security and Intelligence, pp. 93–98 (2010)
16. Zhou, Z., Zhao, G., Pietikäinen, M.: Towards a practical lipreading system. In: Proc. IEEE International Conference on Computer Vision and Pattern Recognition, pp. 137–144 (2011)

Part V
LBP in Various Computer Vision Applications

Chapter 13
LBP in Different Applications

During the past few years the popularity of the LBP approach in various computer vision problems and applications has further increased. In this chapter some representative examples from different areas are briefly described. For a bibliography of LBP-related research and links to many papers, see www.cse.oulu.fi/MVG/LBP_Bibliography.

13.1 Detection and Tracking of Objects

Due to its computational simplicity and discriminative power, LBP has become popular in various object detection and tracking tasks.

Zhang et al. investigated *object detection* using spatial histogram features [51]. The method selects automatically informative spatial histogram features. A hierarchical classifier is learned by combining cascade histogram matching and a support vector machine to detect objects. The spatial histograms are obtained by processing images with a 3×3 LBP operator, and then spatial templates are used to encode spatial feature histograms in scale space. The method was applied to two types of objects: side-view cars from the UIUC image database and text in video frames. High object detection rates are obtained with quite a small number of false detections.

Mu et al. developed discriminative LBPs for *human detection* in personal album [30]. They found that the original LBP does not suit so well for this problem due to its relatively high complexity and lack of semantic consistency. Therefore they proposed two variants of LBP, Semantic-LBP (S-LBP) and Fourier-LBP (F-LBP), see Sect. 2.9. Extensive experiments using the INRIA human database [6] show that the proposed local patterns, especially S-LBP, outperform other gradient-based features. Later, Wang et al. proposed a very effective HOG-LBP method for human detection with partial occlusion handling [46], combining the strengths of the HOG method based on histograms of oriented gradients [6] and LBP.

Grabner and Bischof introduced an *on-line algorithm for feature selection* based on AdaBoost learning [8]. Training the classifier on-line and incrementally as new

M. Pietikäinen et al., *Computer Vision Using Local Binary Patterns*,
Computational Imaging and Vision 40,
DOI 10.1007/978-0-85729-748-8_13, © Springer-Verlag London Limited 2011

data becomes available has many advantages in many applications of computer vision. As features they used Haar-like features, orientation histograms and a simple 4-neighborhood version of the LBP operator. Integral images and integral histograms were used as efficient data structures, allowing a very fast calculation of all these features. A real-time operation was demonstrated in problems dealing with background modeling, tracking, and active learning for object detection. Later they adopted this methodology for *car detection from aerial images* [9].

Ning et al. proposed a robust *object tracking* method using joint color-texture histogram to represent the target and then applying the mean shift algorithm [33]. The major rotation-invariant uniform LBP patterns representing edges, line ends and corners are used to form a mask for joint color-texture feature selection. Experimental results show much better tracking accuracy and efficiency with fewer number of iterations than the original mean shift tracking.

13.2 Biometrics

In addition to face and facial expression recognition, the LBP has also been successfully used in many other applications of biometrics, including iris recognition, fingerprint recognition, palmprint recognition, finger vein recognition and gait recognition.

A *hybrid fingerprint matcher* based on local binary patterns was proposed by Nanni and Lumini [31]. The fingerprints to be matched are first aligned using their minutiae, and then the two images are divided into overlapping subwindows. Each subwindow is convolved with a bank of Gabor filters, and then LBP histograms are computed from the convolved images. Experimental results conducted on the four FVC2002 fingerprint databases show that the proposed method performs very favorably compared to the state-of-the-art.

Vein recognition uses vascular patterns inside the human body. These vascular patterns are in general visible with infrared light illuminators. *Finger vein recognition* uses the unique patterns of finger veins to identify individuals at very high accuracy. Lee et al. developed a method for finger vein recognition using minutia-based alignment and local binary pattern-based feature extraction [23]. The finger vein codes obtained using LBP are robust to irregular shading and saturation factors. The use of LBP reduced false rejection error and thus the equal error rate (EER) significantly. The resulting EER was 0.081% with a total processing time of 118.6 ms.

A *touch-less palm print recognition* system was proposed by Ong et al. [34]. A low-resolution web camera is used to capture images of the user's hand at a distance. A novel hand tracking and region of interest operator are used to capture the palm in real time from the video stream. The discriminative palm print features are extracted by applying LBP descriptor on the palm print directional gradient responses. Promising results are obtained in online experiments. With the proposed system the user verification can be done in less than one second.

Shang and Veldhuis proposed to use local absolute binary patterns (LABP) as image preprocessing for *grip-pattern recognition* in smart gun. In a smart gun the rightful user is recognized based on his handpressure pattern [40]. This application is intended to be used by the police, because carrying a gun in public brings considerable risks. The images in the experimental system are provided by a 44 by 44 piezo-resistive pressure sensor. The modified LBP operator called LABP has two important effects on the grip-pattern images. The pressure values in different subareas within the hand part become much more equalized compared to the original image. After LABP the contrast-enhanced hand-pressure pattern can also be discriminated much better from the background. Due to these effects a significant improvement in the verification performance was obtained.

13.3 Eye Localization and Gaze Tracking

Eye localization for the purpose of face matching in low and standard definition image and video content was investigated by Kroon et al. [19]. A probabilistic eye localization method based on multi-scale LBPs was proposed. The entire eye region was used to learn the eye pattern, and thus requiring no clear visibility of the pupils. The method provided superior performance compared to the state-of-the-art methods in terms of accuracy and efficiency. The standard BioID dataset and an own collection of movie and web cam videos were used in experiments.

The direction of the line of sight, i.e. eye gaze, provides information about a person's focus of attention and interest. Lu et al. [27] developed a method for *gaze tracking* by local pattern model (LPM) and support vector regressor. The proposed scheme is non-intrusive, meaning that users are not equipped with any cameras. The LPM is a combination of an improved pixel-pattern-based texture feature (PPBTF) and uniform local binary pattern feature. LPM is used to calculate texture features from the eye images and a new binocular vision scheme is used for detecting the spatial coordinates of the eyes. The LPM features and the spatial coordinates are fed into support vector regressor to match a gaze mapping function, and then to track gaze direction under allowable head movement. State-of-the-art results are reported in experiments.

13.4 Face Recognition in Unconstrained Environments

Recognition of faces in unconstrained environments has been a topic of increasing interest recently. The problem is very challenging due to large lighting and pose variations, low image resolution, compression artifacts, etc. The number of available training images may also be small. The Labeled Faces in the Wild (LFW) database offers a collection of annotated faces taken from news articles on the web [14]. The problem of LFW recognition was studied e.g. by Wolf et al. [48]. They proposed a novel patch-based variant of the LBP descriptor which was able to improve the performance of the LBP descriptor in both multi-option identification and same/not-same classification tasks (see Sect. 2.9). A state-of-the-art performance for this problem was reported. Later, Ruiz-del-Solar et al. carried out a comparative study of face recognition methods that are suitable to work in unconstrained environments [37]. The conclusion was that LBP-based methods are an excellent election if one needs real-time operation as well as high recognition rates.

Wang et al. [47] investigated boosted multi-task learning for face verification with applications in *web image and video search*. Individual bins of local binary patterns, instead of whole histograms, were used as features for learning, yielding significant performance improvements and computation reduction compared to earlier LBP approaches [2, 50]. A novel Multi-Task Learning (MTL) framework called boosted MTL was proposed for face verification with limited training data. The effectiveness of the approach was shown with a large number of celebrity images and videos from the web.

13.5 Visual Inspection

Visual inspection is economically still perhaps the most important application area of machine vision. Inspection systems can be relatively expensive, as long as they provide high added value, and are therefore attractive testing grounds for new technologies. Typical inspection targets include part assemblies in the electronics and car industry, continuous webs such as paper, steel and fabrics, and natural materials such as wooden boards and coffee beans. Many of these targets are textured and colored, such as wood, and the inspection problem is solved best with target specific methods. One of the major problems is the non-uniformness of real-world textures. Among the first application areas considered for LBP were metal inspection [36] and wood inspection [20, 41].

Turtinen et al. developed a non-supervised method for *paper characterization* [44]. Multi-scale LBP features are extracted from gray scale images, and then the dimensionality of the feature data is reduced to a two-dimensional space with self-organizing map (SOM). With this a self-intuitive user interface and a synthetic view to the inspected data is obtained. The user can select the decision boundaries for different paper classes using the visualized SOM map. After this the SOM is used as a classifier in the testing phase. An excellent classification accuracy of over 99% is obtained in discriminating four different paper quality classes. For these reasons the proposed approach has much potential for on-line paper inspection applications. Later a real-time solution for the same problem was reported [29], utilizing a highly optimized software implementation of the LBP operator, feature reduction, and fast classification.

A method for *separating black walnut meat from shell* using back light illumination was proposed by Jin et al. [16]. Images of walnut meat and shells have different texture patterns due to their different light transmittance properties. The complementary operators, rotation-invariant LBP and gray scale variance, were used for texture description, and a supervised SOM was used as the classifier and for the visualization of multidimensional feature data. An overall separation accuracy of 98.2% was obtained, making the proposed approach to have great potential in walnut processing industry.

Defect detection is very important in fabric quality control. Human inspection of mass products like textiles is expensive and subject to errors. Tajeripour et al. developed a method for fabric defect detection using multiscale LBPs [43].

13.6 Biomedical Applications

The use of LBP in biomedical applications has been recently increasing rapidly. Examples of these developments include:

Image analysis methods that efficiently quantify, distinguish and classify sub-cellular images are of great importance in automated *cell phenotype classification.* Nanni and Lumini [32] developed a reliable method for the classification of protein sub-cellular localization images. In experiments with three image datasets their method based on rotation-invariant LBP features performed better than other well-known methods for feature extraction. Another advantage of the proposed approach is that it does not require cropping of the cells before classification.

Histological tissue analysis can be used for the *diagnosis of renal cell carcinoma* (RCC), requiring exact counts of cancerous cell nuclei. RCC is among the ten most frequent malignancies in Western societies. Fuchs et al. proposed a completely automated pipeline for prediction the survival of RCC patients based on the analysis of immunohistochemical staining of MIB-1 on tissue microarrays [7]. Local binary patterns and color descriptors are used as features, and a random forest classifier detects cell nuclei of cancerous cells and predicts their staining. The system was able to achieve the same superior survival prediction accuracy of renal cell cancer patients as trained medical experts. Local binary patterns have also been applied in histopathological image analysis in supervised image segmentation [39] and tumor morphology based cancer outcome prediction [18].

Li and Meng studied *ulcer detection* in capsule endoscope (CE) images [24]. Capsule endoscopy has wide potential in hospitals, because the entire small bowel can be viewed without invasiveness. A problem is that CE produces too many images and thus a huge burden for physicians. A texture extraction method was proposed for ulcer region discrimination in CE images. The method combines merits of curvelet transform and uniform LBPs, providing an effective description of textures with multi-directional characteristic and robustness to illumination changes. A promising accuracy of over 90% is obtained in experiments.

An approach for *mass false positive reduction in mammographic images* was proposed by Llado et al. [26]. Mammography is the key screening tool for the detection of breast abnormalities from images. The current methods proposed for automatic mass detection suffer from a high number of false positives. A new method for representing the textural properties of masses was proposed, in which the region of interest image is divided into regions from which LBP feature distributions are computed and concatenated into a spatially enhanced descriptor. Support vector machines (SVM) are used for classifying the true masses from the ones being normal parenchyma. The results showed that the LBP features are very effective, providing a better performance than existing methods.

Sorensen et al. studied the area of *lung texture analysis in computed tomography (CT) images* [42]. The specific application area was emphysema quantification, but their results should be applicable to other lung disease patterns as well. Local binary patterns were used as texture features, and joint LBP and intensity histograms were used for characterizing regions of interest. Rotation-invariant LBP performed slightly better than rotation-invariant Gaussian Feature Bank (GFB), and seemed to pick up certain microstructures that are more common in smokers than in people who never smoked.

Due to the high number of medical images routinely acquired in the medical centers, automated classification and retrieval of images has become an important research topic. Jeanne et al. [15] investigated automatic *detection of body parts from X-ray images*. Four conventional features types and local binary patterns were compared using SVM for classification. Comprehensive experiments showed that LBPs provide not only very good global accuracy but also good class-specific accuracies with respect to the features used in the literature.

Unay et al. developed a fast and robust region-of-interest *retrieval method for brain magnetic resonance (MR) images* [45]. Taking into account the intensity-related problems in MR, they used two complementary intensity invariant structure features, local binary patterns and Kanade-Lucas-Tomasi feature points. Incorporating spatial context in the features substantially improved accuracy. Comprehensive experiments showed that dominant local binary patterns with spatial context are robust to geometric deformations and intensity variations and have high accuracy and speed even in pathological cases. The proposed method can not only aid the medical expert in disease diagnosis, or be used in scout (localizer) scans for optimization of acquisition parameters, but also support low power handheld devices.

Facial paralysis is the loss of voluntary muscle movement of one side of the face. Most of the existing objective facial palsy grading systems involve the use of markers on the face. He et al. proposed a method for objective grading of facial paralysis using spatiotemporal LBP-TOP features [12]. Multi-scale features are obtained by processing face images with a Gaussian pyramid and then applying LBP operators with fixed R and P on different scales of the image. A block based approach is used to divide the face into regions, from which the motion information in the vertical and horizontal directions and the appearance features are extracted. The symmetry of facial movements is measured by the Resistor-Average Distance between LBP features extracted from the two sides of the face. An SVM classifier is used to provide quantitative evaluation of facial paralysis. Very promising results are obtained in experiments, outperforming those obtained with an earlier optic flow based method.

13.7 Texture and Video Texture Synthesis

Techniques for data hiding onto images provide tools for protecting copyrights or sending secret messages. Otori and Kuriyama [35] proposed an approach for the *synthesis of texture* images for embedding arbitrary data with little aesthetic defect.

Random coating and re-coating were used to improve the quality of the texture image synthesized from the initial painting using LBP. The algorithm focuses on textures that are iteratively generated by learning a texture pattern of an exemplar.

Video texture synthesis has become an important topic in computer vision, which has applications in games, movies and virtual reality, for example. The goal of synthesis is to provide a continuous and infinitely varying stream of images by doing operations on dynamic textures. Guo et al. [10] proposed a frame-feature descriptor accompanied by a similarity measure using the spatiotemporal LBP-TOP descriptor, which considers both the spatial and temporal domains of video sequences; moreover, it combines the local and global description on each spatiotemporal plane. The preliminary results on different types of video textures were very promising. A starting point for this research was that even though the earlier video texture method proposed in [38] provided quite good visual results, it did not explore well enough the temporal correlation among frames.

13.8 Steganography and Image Forensics

The aim of steganographic techniques is to hide the presence of a message or communication itself from an observer. Avcibas et al. [3] developed a technique for *steganalysis of images* which have been subjected to embedding by steganographic algorithms. The seventh and eight bit planes in an image are used for the computation of several binary similarity measures (BSM). The correlation between the bit planes and the binary texture characteristics within the bit planes will differ between a stego image and a cover image. Local binary patterns were included in the BSM measures used in this method. Simulation results with commercially available steganographic techniques indicated that the proposed steganalyzer is effective in classifying stego and cover images.

The different image processing steps in a digital camera pipeline leave telltale footprints, which can be exploited as forensic signatures. Celiktutan et al. [4] investigated the problem of *identifying source camera of images*, with an aim to develop a method to determine the model and brand of the camera with which an image was acquired. Three sets of forensic features, including binary similarity measures, image quality measures and higher order wavelet statistics, together with SVM classifiers were used to identify the originating camera. Local binary patterns were included in the BSM features as mentioned above. The proposed algorithm worked satisfactorily both for the digital cameras and cell phone cameras.

13.9 Video Analysis

Concept detection plays an important role in video indexing and multimedia retrieval. The aim is to automatically annotate video shots by predefined concept lexicon, i.e. whether a certain concept exists in a video shot or not. The features for concept detection are extracted from the keyframes of each video shot [49]. Le and Satoh presented a framework for efficient and scalable concept detection by fusing SVM classifiers trained by simple features such as color moments, edge orientation histogram and local binary patterns [22]. According to the experiments with various TRECVID datasets, they concluded that due to the LBP feature a higher performance is obtained than with the baseline system which is using Gabor features instead of LBP. The principal goal of the TREC Video Retrieval Evaluation (TRECVID) used is to promote progress in content-based analysis of and retrieval from digital video via open, metrics-based evaluation. Experimental results of Le and Satoh showed that their simple approach can achieve good performance compared to other computationally more complicated systems. An improved approach for concept detection using Markov chain local binary patterns was proposed by Wu et al. [49]. A general framework called Markov stationary features (MSF) was introduced by Li et al. [25] to extend histogram based features. MSF involves spatial structure information of both within histogram bins and between histogram bins. The MSF extension of LBP called MSF-LBP achieved significantly better results that the ordinary LBP in concept detection experiments with TRECVID 2005 and TRECVID 2007 datasets, respectively.

Overlay text provides important semantic clues for video context analysis with applications such as video information retrieval and summary. Most of the earlier methods to extract text from videos are based on low-level features, having problems with varying contrasts or complex backgrounds. A new approach for *detecting and extracting overlay text* from complex video scenes was presented by W. Kim and C. Kim [17]. The method is based on observation that there are transient colors between inserted text and its adjacent background. Local binary patterns are used to describe the texture around transition pixels. Experiments on different types of video show that the proposed approach is robust with respect to changes in character size, position, contrast, and color. It is also language independent.

Crowd estimation is used for crowd monitoring and control in security and video surveillance applications. It is different from pedestrian detection or people counting in the way that no individual pedestrian can be properly segmented in the image. Ma et al. presented a system for crowd density estimation using multi-scale local texture analysis and confidence-based soft classification [28]. A modified block-based version of LBP called Advanced LBP (ALBP) was proposed and adopted as a multi-scale texture descriptor. A weighting mechanism and confidence-based soft classification were used to increase the credibility of the estimations. Experimental results from real crowded scene videos demonstrated the performance and potential of the method. The ALBP features clearly outperformed Gray Level Dependence Matrix and Edge Orientation Histogram features used in earlier crowd estimation studies, and also performed better than the original LBP features.

13.10 Systems for Photo Management and Interactive TV

The popularity of digital cameras and mobile phone cameras has increased rapidly in recent years. Therefore, the sizes of the digital photo albums have grown exponentially. Automatic management of large photo albums has become indispensable. In a photo management system the most challenging task is photo annotation. Cui et al. developed an *interactive photo annotation system* called EasyAlbum [5]. It puts similar faces or photos with similar scene together, and the user can label them in one operation. Contextual re-ranking boosts the labeling productivity by guessing user's intentions. Ad hoc clustering enables users to cluster and annotate freely when exploring and searching in the album, while progressively improving the performance at the same time. In EasyAlbum system local binary patterns are used as facial features, together with color correlogram features extracted from the human body area.

It has been predicted that the future interactive television will provide automatically personalized services for each viewer, such as a personalized electronic program guide, for example. For this purpose the interactive TV should automatically recognize viewers and even their emotions, thus providing feedback about their identities, internal emotions, interests or preferences to the service provider in real-time. Recently, Ho An and Jin Chung proposed an architecture of a *cognitive face analysis system* for future interactive TV [13]. They built a real-time face analysis system containing modules for face detection, face recognition and facial expression recognition. Multi-scale LBP features were computed by scanning the face image with a scalable subwindow. An Ada-LDA learning algorithm was proposed to select the most discriminative LBP features from a large pool of multiscale features generated by shifting and scaling a subwindow over the image. In experiments a good performance was obtained for each of the three tasks using standard sets of test images. A real-time face analysis system including face detection, face recognition and facial expression recognition modules achieved a processing speed of over 15 frames per second. The methods used are, however, too elementary to meet the requirements of a real application environment.

13.11 Embedded Vision Systems and Smart Cameras

Due to its discriminative power and computational efficiency, the LBP method is already being used in many embedded systems, smart cameras, and mobile phones. This section presents some examples of embedded LBP-based systems and smart cameras from the literature.

Computer vision applications for mobile devices are gaining increasing attention due to several practical needs resulting from the popularity of digital cameras in today's mobile phones. For instance, there is a need to develop new technologies to secure the access and the use of services on mobile devices, e.g. through biometric identity verification. The main problem facing the development of computer vision applications for mobile phones concerns the limited memory and CPU resources. Exploiting the low computational cost of LBP, Hadid et al. developed a face authentication prototype for person authentication in mobile phones using Haar-like and LBP features, yielding quite promising results [11]. The system runs at about two frames per second on a Nokia N90 mobile phone with an ARM9 processor with 220 MHz.

The LBP method has also played an important role in a European Commission funded project called MOBIO (www.mobioproject.org) during the period 2008–2010. The main objective of MOBIO project has been to develop robust joint bi-modal (face and speech) authentication on mobile devices. The system was success-fully implemented on the NOKIA N900 mobile phone.

In [21], Lahdenoja et al. proposed a dedicated chip for computing local binary patterns and performing face recognition with a massively parallel hardware, es-pecially with cellular nonlinear network-universal machine (CNN-UM). The face recognition system has the advantage of a speed increase up to 5 times compared to a modern standard computer based implementation, but at the cost of some decrease in LBP flexibility in parameters selection. Zolynski et al. proposed a reformulation of LBP that is efficiently executed on a consumer-grade graphical unit (GPU) [52]. The new implementation is integrated into a pipeline framework that handles the low level data flow between different GPU program elements. Experiments using three types of graphic cards showed a 14 to 18-fold run time reduction compared to standard CPU implementation.

Abbo et al. [1] studied the scalability of LBP based facial expression recognition systems on low-power wireless smart camera platforms. The objective was to iden-tify proper partitioning of the LBP computations over all the resources available on the camera node in order to optimize overall power dissipation. Experiments on a platform with a massively-parallel single-instruction multiple-data (SIMD) proces-sor, showed that the calculation of the LBP labels can be highly optimized by pixel parallel operations while the LBP histogram calculations cannot, thus indicating the sequential nature of the histogram calculation process.

References

1. Abbo, A.A., Jeanne, V., Ouwerkerk, M., Shan, C., Braspenning, R., Ganesh, A., Corporaal, H.: Mapping facial expression recognition algorithms on a low-power smart camera. In: Proc. ACM/IEEE International Conference on Distributed Smart Cameras, pp. 1–7 (2008)
2. Ahonen, T., Hadid, A., Pietikäinen, M.: Face description with local binary patterns: Applica-tion to face recognition. IEEE Trans. Pattern Anal. Mach. Intell. **28**(12), 2037–2041 (2006)
3. Avcibas, I., Kharrazi, M., Memon, N., Sankur, B.: Image steganalysis with binary similarity measures. EURASIP J. Appl. Signal Process. **17**, 553–566 (2005)
4. Celiktutan, O., Sankur, B., Avcibas, I.: Blind identification of source cell-phone model. IEEE Trans. Inf. Forensics Secur. **3**, 553–566 (2008)
5. Cui, J., Wen, F., Xiao, R., Tian, Y., Tang, X.: EasyAlbum: An interactive photo annotation system based on face clustering and re-ranking. In: Proc. CM CHI 2007 Conference on Human Factors in Computing Systems, pp. 367–376 (2007)
6. Dalal, N., Triggs, B.: Histograms of oriented gradients for human detection. In: Proc. IEEE Conference on Computer Vision and Pattern Recognition, vol. 2, pp. 886–893 (2005)
7. Fuchs, T.J., Wild, P.J., Moch, H., Buhmann, J.M.: Computational pathology analysis of tissue microarrays predicts survival of renal cell carcinoma patients. In: Proc. International Confer-ence on Medical Image Computing and Computer Assisted Intervention, pp. 1–8 (2008)

8. Grabner, H., Bischof, H.: On-line boosting and vision. In: Proc. IEEE Conference on Computer Vision and Pattern Recognition, vol. 1, pp. 260–267 (2006)

9. Grabner, H., Nguyen, T.T., Gruber, B., Bischof, H.: On-line boosting-based car detection from aerial images. ISPRS J. Photogramm. Remote Sens. **63**(3), 382–396 (2008)

10. Guo, Y., Zhao, G., Chen, J., Pietikäinen, M., Xu, Z.: Dynamic texture synthesis using a spatial temporal descriptor. In: Proc. IEEE International Conference on Image Processing, pp. 2277–2280 (2009)

11. Hadid, A., Heikkilä, J.Y., Silven, O., Pietikäinen, M.: Face and eye detection for person authentication in mobile phones. In: Proc. ACM/IEEE International Conference on Distributed Smart Cameras, pp. 101–108 (2007)

12. He, S., Soraghan, J.J., O'Reilly, B.F.: Quantitative analysis of facial paralysis using local binary patterns in biomedical videos. IEEE Trans. Biomed. Eng. **56**, 1864–1870 (2009)

13. Ho An, K., Jin Chung, M.: Cognitive face analysis system for future interactive TV. IEEE Trans. Consum. Electron. **55**(4), 2271–2279 (2009)

14. Huang, G.B., Ramesh, M., Berg, T., Learned-Miller, E.: Labeled faces in the wild: A database for studying face recognition in unconstrained environments. Technical Report 07-49, University of Massachusetts, Amherst, 2007

15. Jeanne, V., Unay, D., Jacquet, V.: Automatic detection of body parts in X-ray images. In: Proc. IEEE Workshop on Mathematical Methods in Biomedical Image Analysis, pp. 25–30 (2009)

16. Jin, F., Qin, L., Jiang, L., Zhu, B., Tao, Y.: Novel separation method of black walnut meat from shell using invariant features and a supervised self-organizing map. J. Food Eng. **88**, 75–85 (2008)

17. Kim, W., Kim, C.: A new approach for overlay text detection and extraction from complex video scene. IEEE Trans. Image Process. **18**(2), 401–411 (2009)

18. Konsti, J., Ahonen, T., Lundin, M., Joensuu, H., Pietikäinen, M., Lundin, J.: Texture classifiers for breast cancer outcome prediction. Virchows Arch. **455**(Supplement 1), S34 (2009)

19. Kroon, B., Maas, S., Boughorbel, S., Hanjalic, A.: Eye localization in low and standard definition content with application to face matching. Comput. Vis. Image Underst. **113**(8), 921–933 (2009)

20. Kyllönen, J., Pietikäinen, M.: Visual inspection of parquet slabs by combining color and texture. In: Proc. IAPR Workshop on Machine Vision Applications, pp. 187–192 (2000)

21. Lahdenoja, O., Laiho, M., Maunu, J., Paasio, A.: A massively parallel face recognition system. EURASIP J. Embed. Syst. **2007**(1), 31 (2007)

22. Le, D.-D., Satoh, S.: Efficient concept detection by fusing simple visual features. In: Proc. ACM Symposium on Advanced Computing, pp. 1839–1840 (2009)

23. Lee, E.C., Lee, H.C., Park, K.R.: Finger vein recognition using minutia-based alignment and local binary pattern-based feature extraction. Int. J. Imaging Syst. Technol. **19**(3), 179–186 (2009)

24. Li, B., Meng, M.Q.-H.: Texture analysis for ulcer detection in capsule endoscopy images. Image Vis. Comput. **27**, 1336–1342 (2009)

25. Li, J., Wu, W., Wang, T., Zhang, Y.: One step beyond histograms: Image representation using Markov stationary features. In: Proc. IEEE Conference on Computer Vision and Pattern Recognition, pp. 1–8 (2008)

26. Llado, X., Oliver, A., Freixenet, J., Marti, R., Marti, J.: A textural approach for mass false positive reduction in mammography. Comput. Med. Imaging Graph. **33**, 415–422 (2009)

27. Lu, H.S., Fang, C.-L., Wang, C., Chen, I.-W.: A novel method for gaze tracking by local pattern model and support vector regressor. Signal Process. **90**, 1290–1299 (2010)

28. Ma, W., Huang, L., Liu, C.: Crowd estimation using multi-scale local texture analysis and confidence-based soft classification. In: Proc. Second International Symposium on Intelligent Information Technology Applications, pp. 142–146 (2008)

29. Mäenpää, T., Pietikäinen, M.: Real-time surface inspection by texture. Real-Time Imaging **9**, 289–296 (2003)

30. Mu, Y.D., Yan, S.C., Liu, Y., Huang, T., Zhou, B.F.: Discriminative local binary patterns for human detection in personal album. In: Proc. IEEE Conference on Computer Vision and Pattern Recognition, pp. 1–8 (2008)

31. Nanni, L., Lumini, A.: Local binary patterns for a hybrid fingerprint matcher. Pattern Recognit. **41**, 3461–3466 (2008)

32. Nanni, L., Lumini, A.: A reliable method for cell phenotype image classification. Artif. Intell. Med. **43**, 87–97 (2008)

33. Ning, J., Zhang, L., Zhang, D., Wu, C.: Robust object tracking using joint color-texture histogram. Int. J. Pattern Recognit. Artif. Intell. **23**(7), 1245–1263 (2009)

34. Ong, M.G.K., Connie, T., Teoh, A.B.J.: Touch-less palm print biometrics: Novel design and implementation. Image Vis. Comput. **26**, 1551–1560 (2008)

35. Otori, H., Kuriyama, S.: Data-embeddable texture synthesis. In: Proc. Seventh International Symposium on Smart Graphics, pp. 146–157 (2007)

36. Pietikäinen, M., Ojala, T., Nisula, J., Heikkinen, J.: Experiments with two industrial problems using texture classification based on feature distributions. In: Proc. SPIE Intelligent Robots and Computer Vision XII: 3D Vision, Product Inspection, and Active Vision. Proc. SPIE, vol. 2354, pp. 197–204 (1994)

37. Ruiz-del-Solar, J., Verschae, R., Correa, M.: Recognition of faces in unconstrained environments: A comparative study. EURASIP J. Adv. Signal Process. **2009**, 1–19 (2009)

38. Schödl, A., Szelinski, R., Salesin, D., Essa, I.: Video textures. In: Proc. ACM SIGGRAPH, pp. 489–498 (2000)

39. Sertel, O., Kong, J., Shimada, H., Çatalyürek, Ü.V., Saltz, J.H., Gurcan, M.N.: Computer-aided prognosis of neuroblastoma on whole-slide images: Classification of stromal development. Pattern Recognit. **42**(6), 1093–1103 (2009)

40. Shang, X., Veldhuis, R.N.J.: Local absolute binary patterns as image preprocessing for grip-pattern recognition in smart gun. In: Proc. IEEE International Conference on Biometrics: Theory, Applications and Systems, pp. 1–6 (2007)

41. Silven, O., Niskanen, M., Kauppinen, H.: Wood inspection with non-supervised clustering. Mach. Vis. Appl. **13**, 275–285 (2003)

42. Sorensen, L., Shaker, S.B., de Brujine, M.: Quantitative analysis of pulmonary emphysema using local binary patterns. IEEE Trans. Med. Imaging **29**, 559–569 (2010)

43. Tajeripour, F., Kabir, E., Sheikhi, A.: Fabric defect detection using modified local binary patterns. EURASIP J. Adv. Signal Process. **88**, 12 (2008)

44. Turtinen, M., Pietikäinen, M., Silven, O., Mäenpää, T., Niskanen, M.: Paper characterisation by texture using visualization-based training. Int. J. Adv. Manuf. Technol. **22**, 890–898 (2003)

45. Unay, D., Ekin, A., Jasinschi, R.: Local structure-based region-of-interest retrieval in brain MR images. IEEE Trans. Inf. Technol. Biomed. **14**, 897–903 (2010)

46. Wang, X., Han, T.X., Yan, S.: An HOG-LBP human detector with partial occlusion handling. In: Proc. International Conference on Computer Vision, pp. 32–39 (2009)

47. Wang, X., Zhang, C., Zhang, Z.: Boosted multi-task learning for face verification with applications to web image and video search. In: Proc. IEEE Conference on Computer Vision and Pattern Recognition, pp. 142–149 (2009)

48. Wolf, L., Hassner, T., Taigman, Y.: Descriptor based methods in the wild. In: Proc. ECCV Workshop on Faces in Real-Life Images, pp. 1–14 (2008)

49. Wu, W., Li, J., Wang, T., Zhang, Y.: Markov chain local binary pattern and its application to video concept detection. In: Proc. IEEE International Conference on Image Processing, pp. 2524–2527 (2008)

50. Zhang, G., Huang, X., Li, S.Z., Wang, Y., Wu, X.: Boosting local binary pattern LBP-based face recognition. In: Proc. Advances in Biometric Person Authentication: 5th Chinese Conference on Biometric Recognition, pp. 179–186 (2004)

51. Zhang, H., Gao, W., Chen, X., Zhao, D.: Object detection using spatial histogram features. Image Vis. Comput. **24**(4), 327–341 (2006)

52. Zolynski, G., Braun, T., Berns, K.: Local binary pattern based texture analysis in real time using a graphics processing unit (long version). In: Proceedings of Robotik 2008, VDI-Berichte, vol. 2012. VDI Wissensforum GmbH, Munich (2008)

ERRATUM

Computer Vision Using Local Binary Patterns

Matti Pietikäinen, Abdenour Hadid, Guoying Zhao, and Timo Ahonen

Erratum to: *Local Binary Patterns for Still Images*, **pp. 13–47**
DOI 10.1007/978-0-85729-748-8_2

The following reference:

89. Zhao, G., Ahonen, T., Matas, J., Pietikäinen, M.: Rotation invariant image and video description with local binary pattern features. Under review (2011)

should be updated as follows:

89. Zhao, G., Ahonen, T., Matas, J., Pietikäinen, M.: Rotation invariant image and video description with local binary pattern features. IEEE Trans. Image Process. (in press)

The online version of the original chapter: Local Binary Patterns for Still Images can be found at
http://dx.doi.org/10.1007/978-0-85729-748-8_2.

The online version of the original chapter: Spatiotemporal LBP can be found at
http://dx.doi.org/10.1007/978-0-85729-748-8_3.

The online version of the original chapter: Recognition and Segmentation of Dynamic Textures
can be found at http://dx.doi.org/10.1007/978-0-85729-748-8_7.

Erratum to: *Spatiotemporal LBP,* **pp. 49–65**
DOI 10.1007/978-0-85729-748-8_3

The following reference:

18. Zhao, G., Ahonen, T., Matas, J., Pietikäinen, M.: Rotation invariant image and video description with local binary pattern features. Under review (2011)

should be updated as follows:

18. Zhao, G., Ahonen, T., Matas, J., Pietikäinen, M.: Rotation invariant image and video description with local binary pattern features. IEEE Trans. Image Process. (in press)

Erratum to: *Recognition and Segmentation of Dynamic Textures,* **pp. 109–125**
DOI 10.1007/978-0-85729-748-8_7

1. The original version of the following reference is incorrect:

   9. Doretto, G., Chiuso, A., Wu, Y., Soatto, S.: Dynamic texture segmentation. In: Proc. International Conference on Computer Vision, pp. 1236–1242 (2003)

   The correct reference is as follows:

   9. Doretto G., Cremers D., Favaro P., Soatto, S.: Dynamic texture segmentation. In: Proc. of the Ninth Intl. Conf. on Comp. Vision, pp. 1236–1242 (2003)

2. The citation should be "Doretto et al. (2003)" not "Doretto (2003)".

3. The following reference:

   27. Zhao, G., Ahonen, T., Matas, J., Pietikäinen, M.: Rotation invariant image and video description with local binary pattern features. Under review (2011)

   should be updated as follows:

   27. Zhao, G., Ahonen, T., Matas, J., Pietikäinen, M.: Rotation invariant image and video description with local binary pattern features. IEEE Trans. Image Process. (in press)

Index